POLYMER SCIENCE AND TECHNOLOGY

HYDROGELS IN BIOLOGY AND MEDICINE

POLYMER SCIENCE AND TECHNOLOGY

Additional books in this series can be found on Nova's website under the Series tab.

Additional E-books in this series can be found on Nova's website under the E-books tab.

Polymer Science and Technology

Hydrogels in Biology and Medicine

J. Michalek
M. Pradny
K. Dusek
M. Duskova
R. Hobzova
and
J. Sirc

Nova Science Publishers, Inc.
New York

Copyright © 2010 by Nova Science Publishers, Inc.

All rights reserved. No part of this book may be reproduced, stored in a retrieval system or transmitted in any form or by any means: electronic, electrostatic, magnetic, tape, mechanical photocopying, recording or otherwise without the written permission of the Publisher.

For permission to use material from this book please contact us:
Telephone 631-231-7269; Fax 631-231-8175
Web Site: http://www.novapublishers.com

NOTICE TO THE READER
The Publisher has taken reasonable care in the preparation of this book, but makes no expressed or implied warranty of any kind and assumes no responsibility for any errors or omissions. No liability is assumed for incidental or consequential damages in connection with or arising out of information contained in this book. The Publisher shall not be liable for any special, consequential, or exemplary damages resulting, in whole or in part, from the readers' use of, or reliance upon, this material.

Independent verification should be sought for any data, advice or recommendations contained in this book. In addition, no responsibility is assumed by the publisher for any injury and/or damage to persons or property arising from any methods, products, instructions, ideas or otherwise contained in this publication.

This publication is designed to provide accurate and authoritative information with regard to the subject matter covered herein. It is sold with the clear understanding that the Publisher is not engaged in rendering legal or any other professional services. If legal or any other expert assistance is required, the services of a competent person should be sought. FROM A DECLARATION OF PARTICIPANTS JOINTLY ADOPTED BY A COMMITTEE OF THE AMERICAN BAR ASSOCIATION AND A COMMITTEE OF PUBLISHERS.

Library of Congress Cataloging-in-Publication Data
Hydrogels in biology and medicine / J. Michalek ... [et al.].
   p. ; cm.
Includes bibliographical references and index.
ISBN 978-1-61668-758-8 (softcover)
1. Colloids in medicine. I. Michalek, J.
[DNLM: 1. Hydrogels. 2. Biocompatible Materials. QT 37.5.P7]
R857.C66H93 2010
610.28--dc22
                                                 2010025384

Published by Nova Science Publishers, Inc. ✤ *New York*

Contents

Preface		vii
Chapter 1	Introduction	1
Chapter 2	Structure of Hydrogels in Relation to the Formation Conditions	3
Chapter 3	Swelling and Mechanical Properties of Hydrogels	15
Chapter 4	Contact Lenses	41
Chapter 5	Intraocular Lenses	49
Chapter 6	Functional Implants	53
Chapter 7	Blood Vessel Embolization	55
Chapter 8	Wound Dressings	57
Chapter 9	Conductive Hydrogels for Biomedical Use	63
Chapter 10	Hydrogels in Tissue Engineering	67
References		71
Index		85

Preface

The range of materials used for biomedical applications is very broad. This means that the demands on their properties are very diverse depending on various medical areas and applications. Moreover, it is often necessary to have available materials with the possibility to set the required parameters very precisely in very wide ranges. Because of the similar mechanical behaviour of hydrogels with that of living tissues and their good compatibility and ability of hydrogels to swell in water, the hydrogels are often used in biomedical applications.

Hydrogel polymers are natural or synthetic hydrophilic crosslinked polymers favourably interacting with water. The swelling degree (the water content in equilibrium-swollen gel) is a function of polymer hydrophilicity and the degree of crosslinking and has influence on a number of physical and chemical properties (e.g., refractive index, transport properties). In biomedical applications, the physical structure and morphology of the hydrogels is adjusted to the targeted performance and both homogenous and heterogeneous materials with porous, nanofiber, or nanoparticle structures are used.

One of the oldest and still widely used biomedical applications of hydrogels is the contact lens. Hydrogels are successfully applied as a variety of implants in surgery, ophthalmology, otorhinolaryngology, neurology, urology, gynaecology etc. In addition, hydrogels are used to cover wounds, burns, trophic defects, or as a two-dimensional supports for cultivation and potential transplantation of cells or as three-dimensional scaffolds for tissue engineering and cell therapy.

In some applications, particularly, in tissue engineering, the biodegradable materials (hydrogels) are used. The advantage is that after the fulfilment of their tasks (e.g., proliferation of cell culture), they disintegrate (hydrolytically or enzymatically) and, subsequently, they are eliminated from the organism. Controlled transport and release of drugs is a separate biomedical area.

Hydrogels for biomedical applications can be generally classified in different categories depending on their interaction with the living tissue. The hydrogels are used either in direct contact with tissues (implants, injection needle, infusion sets, bandages, suture materials, carrier of drugs, cell or tissue culture) or in indirect contact with tissues (orthesis, medical devices, air filters, sanitary supplies, etc.).

Chapter 1

Introduction

No exact definition of the "Gel" or "Hydrogel" state exists. It is commonly accepted that the gel is a viscoelastic solid containing a certain amount of liquid. The viscoelastic characteristics vary in a certain range corresponding to the softness of the gel and certain shape reversibility when external stresses are applied (cf., e.g., [1]). The Encyclopedia of Polymer Science offers the following definition for *hydrogel* "*Hydrogels* are hydrophilic polymers that absorb water and are insoluble in water at physiologic temperature, pH, and ionic strength because of the presence of a three-dimensional network" [2]. Since we will be discussing here polymer gels we will be using the latter definitions but we have to pay attention to the nature of bonds by which three-dimensional network structure is achieved. As far as "hydrogels" concerns, these are "gels" which favourably interact with water and in which water is the predominant medium during hydrogel service.

The bonds making from a polymer the gel are either covalent (permanent) or physical (transient). The covalent bonds are strong and durable. The majority of them does not dissociate, they are rather split or transformed by chemical reactions. The range of relaxation time (durability) of bonds is very broad. It falls down from years for covalent bonds to seconds for entangled polymer solutions. Some physical gels exhibit the behaviour of covalent gels although the isolated individual bonds making them physically crosslinked are weak (hydrophobic interactions, hydrogen bonds). However, they are formed cooperatively and are grouped in sequences of various lengths (crystallites, multiple helices, etc., cf., e.g., [3]). The difference between covalent and sequential physical bonds exists in their temperature response. The covalent bonds usually do not dissociate upon increasing temperature until they degrade at temperatures exceeding the boiling point of water. A relatively sharp dissociation or "melting" temperature is characteristic for physical hydrogels. This dissociation is manifested by transition of the gel to solution or an abrupt change in the degree of swelling (cf., e.g., [4]). For hydrogels, the sol-to-gel transition often occurs on lowering the temperature.

In this chapter, we will concentrate on covalent gels. In hydrogels, the physical interactions always play an important role.

Chapter 2

Structure of Hydrogels in Relation to the Formation Conditions

2.1. General Features of Network Build-Up

Formation of hydrogels from the starting materials (precursors) by chemical crosslinking as well as physical association has some common features [5],[6]. To form hydrogels, the precursor must have a certain affinity to water. First, the molecular weights of the forming reaction or association products increase. The weight-average (M_w) and higher averages of molecular weight increase much steeper than the number-average molecular weight (M_n), so that the polydispersity index characterized by the ratio M_w/M_n also increases. If the system gels at all, there always exists a point, the gel point, characterized by critical conversion of the functional groups (covalent systems) or degree of association (physical gels). Experimentally, the gel point is described, respectively, by critical reaction time and critical concentration. Just beyond the gel point, the gel fraction (fraction of polymer of „infinite" molecular weight) appears which for covalently crosslinked gels is insoluble in any good solvent. The structural changes also serve for determining the gel point - particularly the divergence of M_w or the conversion at which the gel fraction is extrapolated to zero. Also, some other physical methods indicate the occurrence of the gel point like independence of the loss angle tangent of frequency in dynamic mechanical experiments [7], changes in the structure of laser speckle [8], time-resolved light scattering [9] or microrheology in which reaching of the gel point is manifested by a sudden change of slope of the dependence of the mean-square displacement of markers on time [10]. For the network (gel) formation a difference in the composition of the sol and gel is characteristic. Especially the difference of the degree of conversion (or association degree) of functional groups in the sol and gel is large.

In the region beyond, the gel point is characterized by a continuous transformation of soluble molecules into gel by reaction of the functional groups of sol molecules with the groups of the gel and build-up of the internal structure of the gel. This gel build-up is associated with increasing connectivity of the gel units manifested by increasing number of paths called elastically active network chains (EANC) and decreasing number of dangling chains composed of units connected only by single paths of bonds to the gel. The dangling chains are not active in resisting stresses applied externally [5],[6]. Typical changes characterizing covalent network formation are shown in Figure 1.

The gel point (located in this example at about 20% conversion) can vary from close to 0% conversion of reactive groups (crosslinking of long primary chains) to a value close to 100% conversion depending mainly on the functionality of precursors and stoichiometric disbalance in systems based on alternating reactions.

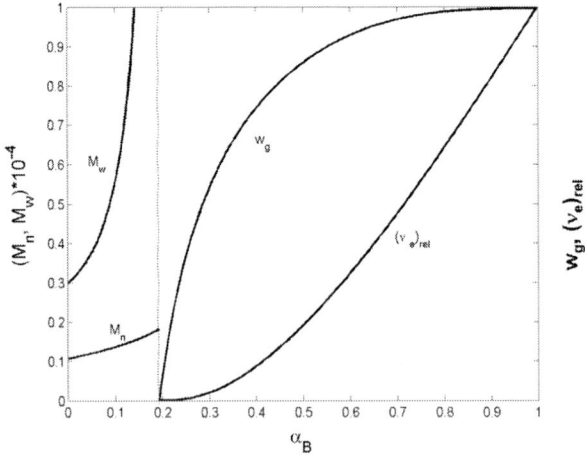

Figure 1. Development of molecular weight averages (M_n, M_w), gel fraction (w_g), and relative concentration of EANC's ((v_e)$_{rel}$) for a statistical copolymer of units with primary and secondary OH groups (2:1) and a non-functional monomer crosslinked with a triisocyanate; α_B is conversion of NCO groups [6].

The dependence of M_w is always curved upwards. The increase of the gel fraction is initially steep and then slowly converges to 100 %, but it is not always so. Sometimes, a part of the system remains soluble despite the conversion reached 100 %. The dependence of concentration of elastically active network chains on conversion is initially curved upwards and may go over to a linear dependence. The shift of the gel point depends on functionality distribution of the precursor, group reactivities, and especially on the reaction mechanism. These issues will be discussed in Section 2.3.

2.2. Physical Polymer Hydrogels

Physical gels can have manifold structures depending on the type of association and concentration. The degree of ordering can be low for randomly distributed associating groups – stickers – or high for system containing ordered sequences (block copolymers, peptide units) which give rise to micellar, cylindrical, or layered lamellar networks (gels). The latter systems are the domain of supramolecular chemistry. Another class of hydrogels are biohybrid gels in which the biopolymer motifs, usually oligopeptides or polypeptides, are chemically attached to the synthetic hydrophilic chains. The gels are formed by association of the motifs and are liquefied by motifs dissociation. The motifs associate to form superhelices or β-sheets. The association is highly specific and dependent on the exact sequence of amino acids in the motifs, which is utilized in molecular recognition

2.3. Inter- and Intramolecular Crosslinking

A common phenomenon which is often overlooked is the formation of cyclic structures by intramolecular reactions. Before the gel point, formation of a bond between two functional groups of one molecule is an intramolecular reaction. In contrast to intermolecular bonds, formation of an intramolecular bond does not contribute to the increase of molecular weight (Figure 2).

The molecular weight averages of a crosslinking system grow more slowly and the gel point is shifted to higher conversions [11] compared to the situation if all bonds were intermolecular.

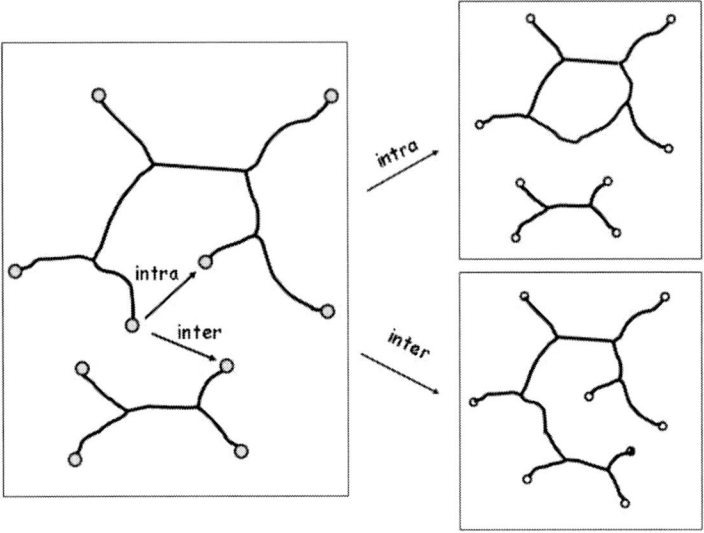

Figure 2. Inter- and intramolecular crosslinking before the gel point.

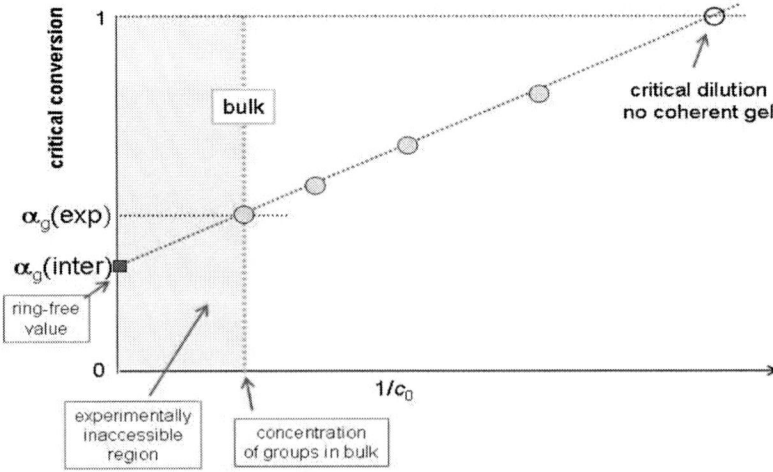

Figure 3. Determination of the effect of cyclization by measurement of the gel point conversion of functional groups, $\alpha_g(\exp)$, in dependence on the reciprocal initial concentration of functional groups $1/c_0$.

Beyond the gel point, some crosslinks are not effective in the increase of the concentration of EANCs. The extent of cyclization increases with dilution of the system. The shift of the gel point conversion in dependence on dilution during network formation characterizes the tendency of the given system to formation of cycles. Moreover, extrapolation of the dependence to the hypothetical infinite concentration of functional groups gives the ring-free value of the gel point conversion (α_g(inter) in Figure 3). This is the value which can be relatively easily obtained theoretically.

For physical gels, gelation is primarily dependent on temperature, especially for bonds formed cooperatively (helices, superhelices, β-sheets). However, in the temperature region where such bonds can exist, gelation is a function of concentration [11]. Because such critical concentrations are usually low, sometimes well below 1 wt.-%, a considerable fraction of bonds is intramolecular [12].

2.4. Ways of Formation of Polymer Hydrogels

This Chapter is not designed for listing various chemistries being used for synthesis and manufacture of hydrogels. It only discusses the significance of the mechanism of chemical network build-up for gelation and network structure. Basically, three ways of network formation exist differing in the mechanism of bond formation:

1. crosslinking of preformed polymer chains
2. chain (co)polymerization
3. step growth reactions

The boundaries are not sharp because preformed chains can be short or long and they may contain groups reacting stepwise or groups (e.g., C=C double bonds) reacting chainwise. The basic difference between step and chain reactions exists in the evolution of molecular weight distribution. In the case of step reactions, the functional group on any oligomer can react with any other corresponding group and the distribution develops from monomer to dimer, trimer, tetramers, etc. In the chain reactions, it is only the activated group that can react with a non-activated one. If the propagation step is relatively fast compared to initiation, the branched chains are initially "dissolved" in monomers. These differences in evolution of molecular weight distribution basically affect the position of the gel point and the onset of possible phase separation.

Comparing as an example crosslinking of a tetrafunctional tetrol with stoichiometric amount of a bifunctional diisocyanate and copolymerization of tetrafunctional glycol dimethacrylate with bifunctional methyl methacrylate, we find that the former, the step system, gels at around 60% conversion, whereas the latter, the chain system, can gel already at conversions of C=C bonds as low as 0.1 %. The corresponding values of the critical conversion (α_g) for the ideal ring-free case are given by relations

$$\alpha_g = [(f_H - 1)(f_I - 1)]^{-1/2} = 3^{-1/2} \qquad \alpha_g = 1/x_4 P_w^0$$

step growth build-up   chain build-up

where f_H and f_I are numbers of functional groups of H-component (OH groups) and I-component (isocyanate groups), respectively, x_4 is the fraction of double bonds of the tetrafunctional monomer and P_w^0 is the weight average degree of polymerization of the polymerized sequences of bonds ("primary chains").

Polymer gels are prepared not only from monomers but frequently also from preformed precursors, such as telechelic polymers, functional stars, linear copolymer chains, hyperbranched polymers, and various other prereacted systems [6]. Often, distributions exist not only in the molecular weight but also in the number of functional groups per molecule and distribution in chemical reactivities. These distributions can affect gel formation and gel structure considerably.

Recently, ways of formation of hydrogels were enriched by easily proceeding reactions in aqueous media, such as the click chemistry (cf., e.g., a compendium [13]).

$$R1-\!\!\equiv\ +\ N\!=\!N\!=\!N\!-\!R2\ \longrightarrow\ \begin{array}{c} R1 \\ \diagup\!\!\diagdown \\ N\!=\!N\diagdown \\ R2 \end{array}$$

Diazirine-based photo-reactive crosslinkers for crosslinking of amine functionalities including living cells [14] are interesting as well.

[Reaction scheme: R1–CH$_2$CH$_2$–NH$_2$ + N-hydroxysuccinimide ester with diazirine group → protein 1–NH–C(=O)–CH$_2$CH$_2$–C(CH$_3$)(N=N) + uv 350n + protein 2, N$_2$ → protein 1–NH–C(=O)–CH$_2$CH$_2$–C(CH$_3$)–protein 2]

Hydrogels can be prepared in the absence of diluents and swollen in water after their preparation. Alternatively, preparation of hydrogels can occur in the presence of a diluent. The diluent is either water, or another additive which is then (possibly in several steps) exchanged for water. By varying the mechanism of network build-up and nature and amount of diluents, manifold morphologies of hydrogels of various chemical natures can be prepared (cf., e.g., [15]).

Traditional and widely used is free-radical polymerization and copolymerization of vinyl monomers containing a hydrophilic group, such as 2-hydroxyethyl methacrylate or acrylate, 2-hydroxyethoxyethyl methacrylate, glycerol monomethacrylate, acrylamide and methacrylamide, substituted acryl- and methacrylamides, monomers containing combination of hydroxyl and amide groups, carboxyl groups of acrylic or methacrylic acid. Some crosslinkers do not have sufficiently high affinity to water. Presence of a hydrophilic group like in glycerol dimethacrylate helps to solve this problem.

For hydrogels prepared by step growth reactions of alternating type, the effect of stoichiometric imbalance can be employed giving excess of hydrophilic groups in the network. Crosslinked polyamides from polyamines and polyacids, or crosslinked polyurethanes with excess of OH groups can serve as examples. The polyurethanes cannot be prepared in water or alcohols because of their reaction with the isocyanate group, but another solvent has to be used which is later exchanged for water. The crosslink density and hydrophilicity of the resulting network can be tuned by varying the off-stoichiometry, the hydrophilicity of components and their functionality (including bifunctional and monofunctional components). To disentangle this maze of effects is a nice task for the branching theory. A prominent component bringing in hydrophilicity are oligomers or low-molecular-weight polymers of ethylene glycol.

Existing chains are in fact high-functionality precursors and can be crosslinked by step growth reactions like poly(vinyl alcohol) (PVA) with glyoxal, or chain reactions like PVA with grafted acrylic groups.

2.5. Special Features of Polymerization in Presence of Diluents. Phase Separation - Macrosyneresis vs. Microsyneresis

It sometimes happens, intentionally or undesirably, that the polymerizing system loses its thermodynamic stability and phase separates. This usually takes place if a diluent (poor or good solvent or other additive like a polymer) is present during polymerization. A two-phase system results – one phase is usually the (swollen) crosslinked polymer and the second phase is composed mainly of uncrosslinked additives. More complicated cases are not excluded – three phases in equilibrium, two crosslinked phases, etc.

The uncrosslinked phase can separate from the polymerizing system as a bulk (liquid) phase or the separated phases are interdispersed. These phase separation processes are called *macrosyneresis* and *microsyneresis*, respectively [6],[16]. The morphology of the gel depends on relative volumes of the separated phases (dispersion of diluent in the crosslinked matrix or dispersion of the crosslinked polymer in the liquid). The morphological features depend on several factors, such as interfacial tension, progress of polymerization at incipient phase separation – how far the system is from the gel point, how dense the network is, etc.

To understand the changes occurring during gel formation, one should examine the effect of swelling changes induced by such change in temperature by which the gel deswells. The result depends on crosslink density. If the crosslink density is relatively high, the volume of the piece of gel decreases and the gel remains transparent (homogeneous). If the crosslink density is low, turbidity appears first and the volume of the gel very slowly decreases. The final result after long time (months) [3],[16] is the same as with macrosyneresis - two bulk phases (Figure 4).

Microsyneresis causes first separation of the excess solvent in the form of "droplets" inside the gel and the droplets are under pressure of the locally deformed network.

Structure of Hydrogels in Relation to the Formation Conditions 11

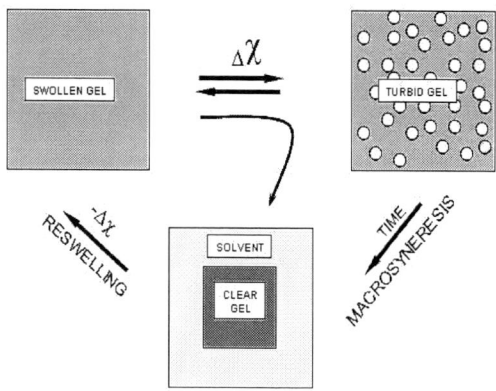

Figure 4. Changes induced in a gel by change of the interaction parameter χ (e.g., by change of temperature). Microsyneresis goes over to macrosyneresis.

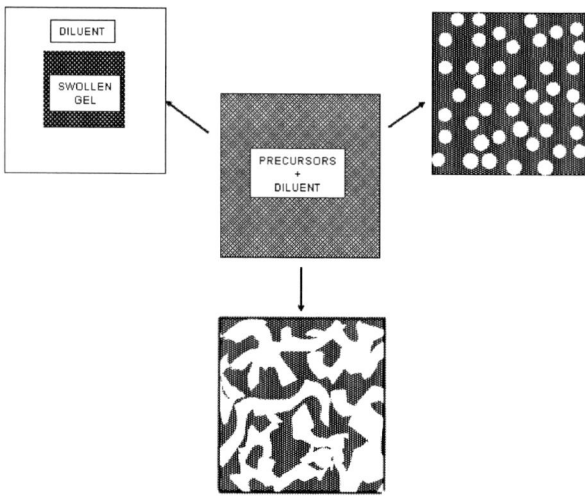

Figure 5. Phase separation occurring during network formation. Macrosyneresis and two forms of microsyneresis.

For low crosslink density, this pressure and the associated driving force for expulsion of the droplets are very low. In the case of reaction (polymerization) induced phase separation, the situation is similar. When phase separation occurs close to the gel point, where the effective crosslink density is low, phase separation always occurs by microsyneresis and the heterogeneous structure is fixed by proceeding crosslinking. The heterogeneous structure can never relax to two bulk phases. It is usually so when porous (macroporous) gels are prepared by adding diluents to the monomers. In some other cases, when the network is strong enough, macrosyneresis is the dominant phase separation mechanism [17] (Figure 5).

The reason for the phase separation is simple: the system cannot tolerate that much diluent it contains. The intolerance appears due to increasing crosslink density (concentration of EANCs, v_e). This is called v-syneresis. Deteriorating polymer-diluent interaction characterized by increasing polymer-solvent interaction parameter χ results in χ-syneresis. Sometimes, both factors – increasing v_e and increasing χ - are operative (cf., e.g., [17]). The effects of conversion degree at incipient phase separation and phase-volume ratios developed afterwards can be quantified using the swelling theory as will be shown in the Section 3.1.

Many hydrogel-forming systems are based on free-radical copolymerization in the presence of diluent. It should not be forgotten that the monomer(s) are not only parent materials for the hydrogel polymer but also a diluent. When an inert, non-polymerizing diluent is added we should consider the systems as a ternary one (disregarding the small amount of crosslinker) – network polymer-diluent 1 (monomer)-diluent 2 - or at least a pseudobinary system in which the diluent is a mixed solvent whose properties change continuously. For instance, HEMA with water form a cosolvent mixture for the polyHEMA [18]. PolyHEMA gels swell better in the monomer than in water and the maximum degree of swelling is located at about 50 wt.-% water.

Figure 7 shows that the critical conversion at which phase separation starts decreases with increasing water content which also means that the pore volume as well as their connectivity increase. The cosolvency of water-HEMA mixture causes in fact a shift of the critical conversion to higher values compared to the situation when water and HEMA would interact specifically. These studies have further shown some other specific features of this mainly χ-driven phase separation: (a) the degree of swelling of the gel phase does not change much after phase separation has taken place; (b) in the region of dilutions exceeding the critical value but not too high, microsyneresis can be accompanied by macrosyneresis; (c) depending on the polymerization rate, reaction rate can exceed the rate of phase separation.

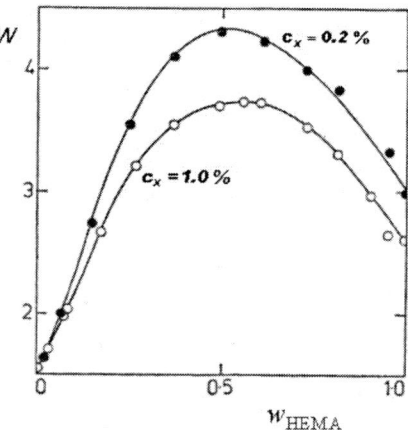

Figure 6. Equilibrium degree of swelling of polyHEMA gel in mixtures of HEMA with water. W = swollen weight/dry weight; c_x concentration of ethylene dimethacrylate crosslinker in wt.-%; w_{HEMA} weight fraction of HEMA in the swelling liquid; 50 °C; data of ref. [18]

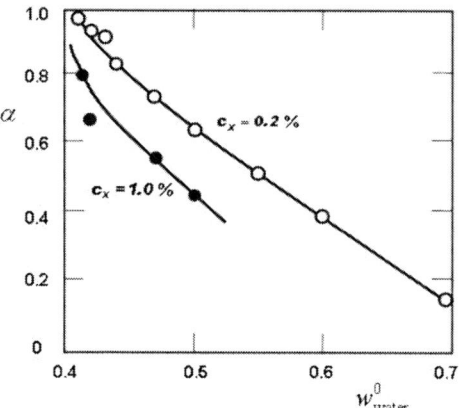

Figure 7. Conversion of double bonds of HEMA, α, in HEMA-water mixtures at incipient phase separation; w^0_{water} is the initial weight fraction of water in HEMA-water mixtures; c_x concentration of ethylene dimethacrylate crosslinker in wt.-%; 60 °C; data of ref. [18].

Although the poor thermodynamic quality of water is the main driving force for phase separation, the concentration of crosslinker is also important. For swelling in water the increase of concentration of ethylene dimethacrylate (EDMA) from 0.2 to 1.0 wt.-% does not play any role, in the better cosolvent system of water with the monomer the difference in swelling degree is appreciable and because of that also the critical conversions differ (Figure 6 and Figure 7).

Chapter 3

Swelling and Mechanical Properties of Hydrogels

3.1. Swelling Properties of Hydrogels

Swelling is the most characteristic property of a gel and swelling in water is characteristic for a hydrogel. Here, we will consider hydrogels that take up water (or other solvent) to a limited extent, i.e. they swell to equilibrium. The equilibrium degree of swelling at a given temperature and solvent vapour pressure is a characteristic property of a hydrogel. For several applications, also the swelling dynamics is important. However, the swelling and deswelling rates depend on sample geometry and on the initial state of the material (rubbery or glassy).

In this chapter, we will focus on swelling equilibria. In addition to weak interactions including hydrophobic interaction and hydrogen bonding, we will consider the effect of charged groups, conditions for a three-phase equilibrium, and briefly ternary systems consisting of polymer-solvent1-solvent2. Based on the swelling thermodynamics, we will formulate the conditions for phase separation.

3.1.1. Equilibrium Swelling

Swelling occurs because polymer segments have tendency to mix with solvent molecules. Mixing is associated with large gain in entropy. In the case of an uncrosslinked polymer, mixing of polymer with is affected by interactions between polymer segments and solvent molecules. At concentrations at which polymer coils overlap, a temporary, dynamic network can be formed having the properties of the gel in which the network chains are relaxed. If the interactions are changed, the physical gel can take up more solvent. It is eventually is transformed into a solution which can be further diluted.

In a covalently crosslinked network, the mixing tendency of polymer segments and solvent molecules is opposed by network connectivity and limited by chain elasticity. As a result of solvent uptake, the elastically active network chains are stretched and the resulting retractive force is counterbalanced the osmotic pressure.

In the thermodynamic treatment, the additivity of the Gibbs energies due to mixing and isotropic deformation of the network is usually assumed

$$\Delta G_{sw} = \Delta G_{mix} + \Delta G_{net} \tag{1}$$

For a binary systems, crosslinked polymer – solvent, the Flory-Huggins polymer solution theory and Flory, or Flory-Erman rubber elasticity theory give the following expression for ΔG_{sw} (cf., refs. [19]-[21])

$$\Delta G_{sw}/kT = n_1 \ln\phi_1 + n_1\chi\phi_2 + n_e\left(A(\Lambda_x^2+\Lambda_y^2+\Lambda_z^2-3) - B\ln\Lambda_x\Lambda_y\Lambda_z\right) \tag{2}$$

where the deformation ratios Λ_i are related to isotropic reference state. For isotropic swelling, $\Lambda_x = \Lambda_y = \Lambda_z = \Lambda$

$$\Delta G_{sw}/kT = n_1 \ln\phi_1 + n_1\chi\phi_2 + 3n_e(A\Lambda^2 - B\ln\Lambda) \tag{2a}$$

where k is Boltzmann constant, T temperature in Kelvin, ϕ_1 and ϕ_2 are volume fractions of solvent and polymer, respectively, n_1 and n_e are the number of solvent molecules and elastically active network chains (EANC), respectively, χ is the Flory-Huggins interaction parameter, A and B are factors in Flory-Erman junction-fluctuation rubber elasticity theory [21]. The deformation ratio for EANCs, Λ, is equal to

$$\Lambda = \phi_2^{-1/3}\phi_0^{1/3} \tag{2b}$$

where ϕ_0 is volume fraction of polymer components at network formation, $1 - \phi_0$ is the fraction of non-polymerizable diluent. The values of the factors A and B depend on how much interchain interactions affect fluctuation of crosslinks. Within the framework of this theory, there are two limits: $A = (f_e - 2)/f_e$, $B = 0$ for a phantom network with freely fluctuating crosslinks, and $A = 1$, $B = 2/f_e$ for suppressed fluctuation of crosslinks (f_e is the number of infinite paths issuing from an elastically active crosslink; in a perfect network it is equal to the chemical functionality).

From ΔG_{sw}, the change of the chemical potential of the solvent is obtained

$$\Delta \mu_1 / RT = \ln a_1 = \ln(1-\phi_2) + \phi_2 + \chi \phi_2^2 + V_1 \nu_e (A\phi_2^{1/3} \phi_0^{2/3} - B\phi_2) \quad (3)$$
$$a_1 \approx p_1 / p_1^0$$

The swollen network is in equilibrium with pure solvent or with solvent vapour of partial pressure p_1, p_1^0 being the vapour pressure of pure solvent. The activity of pure solvent is by definition $a_1 = 1$ and then the right-hand-side of Eq. (3) is equal to zero. If the gel is kept in contact with solvent vapours, it will swell less: the r.h.s of Eq. (3) will by less than zero. Thus knowing the structural parameters (χ, ν_e, ϕ_0, and for the given values of A and B), one can calculate the volume degree of swelling, expressed as $1/\phi_2$ equal to volume of swollen sample/volume of dry sample.

The swelling equilibria are schematically shown in Figure 8 as interdependence of the volume fraction of polymer and temperature for a system having lower (LCST) and upper (UCST) critical solution temperature, compared with an uncrosslinked (linear, monodisperse) polymer of increasing molecular weight.

For uncrosslinked polymers, the coexisting phases have different polymer concentrations, the swollen gel is in equilibrium with pure solvent ($\phi_2 = 0$).

The degree of swelling of a crosslinked polymer is determined by

1. concentration of elastically active network chains, ν_e,
2. polymer solvent interaction parameter χ (or g) and its concentration and temperature dependences,
3. memory parameter ϕ_0 characterizing dilution at network formation and thus the state of coiling of network chains in the dry network,

4. molar volume of the solvent, V_1,
5. functionality of the crosslink (average number of bonds with infinite continuation per elastically active crosslink, f_e,
6. ordering of the swelling liquid (liquid-crystalline solvents),
7. presence of macromolecular substances that lower solvent activity,
8. external strains, if applied.
9.

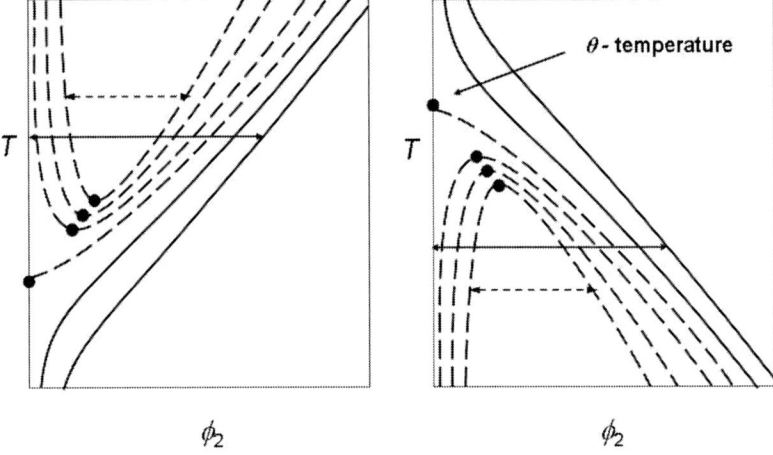

Figure 8. Partially miscible polymer solutions and swollen crosslinked polymers. Dashed curves - binodals for a linear polymer of increasing molecular weight; the last curve for very high (infinite, M_{inf}) molecular weight; • critical points, the last for M_{inf} defines the θ-temperature. Horizontal double arrows define the composition of coexisting phases.

Role of the Interaction Parameter χ

The goodness of solvents increases with decreasing χ. However, χ is usually not a constant but depends on temperature and polymer concentration. The interaction term should be then considered as a excess chemical potential. The temperature and concentration dependences of χ exist for many non-aqueous systems but are quite strong for aqueous gels. Several semiempirical have been offered but it is difficult to obtain the respective parameters independently and not just by curve fitting. The complicated dependence of χ is due to existence of different kinds of interacting sites on the polymer, hydrogen bonding, and iceberg structure of water affected by organic solutes (hydrophobic interactions).

The *temperature dependence* of swelling is positive for UCST systems and negative for LCST systems. For "organic" systems, usually but not always the UCST behaviour prevails; aqueous systems usually show up LCST. However, the temperature dependence is sometimes more complicated, iceland of partial miscibility are encountered. For instance, polyHEMA shows a shallow minimum at about 55 °C (Figure 9). Several functional forms for the dependence of χ on T have been offered in the literature. The most common is the dependence

$$\chi_T = a + \frac{b}{T} \tag{4}$$

corresponding to simple forms of entropic and enthalpic part of the interactions, sometimes the temperature dependence is more complicated (cf., e.g., [23],[24]) which leads to the presence of a logarithmic term

$$\chi_T = \frac{b_1}{T} + \ln\frac{c}{1+dT} \tag{5}$$

However, it is questionable whether the temperature dependence can be factored out as $\chi(T,\phi_2) = \chi_T(T)\chi_c(\phi_2)$, if a concentration dependence of χ exists. For a deeper discussion of the interaction function, see ref. [22].

Very important is the *concentration dependence* of χ.[1] The concentration dependence of χ is more frequent than a concentration independence. It can be generally expressed as power series of ϕ_2 for χ or g. Often, the function

$$\chi_c = \frac{\chi_T}{1-b\phi_2} = \chi_T(1 + b\phi_2 + b^2\phi_2^2 + b^3\phi_2^3 + \cdots) \tag{6}$$

is used quite successfully to express the combined concentration and temperature dependence. However, a strong concentration dependence of χ can give rise to so-called "off-zero critical concentration" [25],[26]. In this

---

1 Equations (2) and (3) are selfconsisten only if χ is concentration independent, since the chemical potential is obtained from Gibbs energy by differentiation.. Most frequently, the concentration dependence is determined by using eq. (3). Then, the concentration dependence of χ in eq. (2) is different. in more recent literature [4], an interaction function g(ϕ2) is used. The interrelations can be found in ref.. [4].

case, the critical concentration for a polymer of infinite molecular weight is not zero. Such a dependence can induce a volume-phase transition (see below). The concentration dependence of χ is usually determined by measuring the the degree of swelling as a function of crosslink density while the concentration of EANCs is determined independently,. In this case, the concentration dependence of χ may also be affected by constraints to interaction imposed by the crosslinks (cf., e.g. ref. [27]). Therefore, it is better to determine the concentration dependence of χ by measuring the solvent uptake at various vapour pressures of the solvent.

In water, the swelling degree of polyHEMA passes through a shallow minimum at about 55 °C, whereas in more organic butanol the swelling degree increases with temperature almost linearly. However, many hydrophilic systems show up a decrease of swelling degree with increasing temperature. The popular polyNIPA (poly(N-isopropylakrylamide)) can serve as example. The concentration dependence of the interaction parameter can be determined only in a very limited range of polymer concentrations because even uncrosslinked polyHEMA does not swell much. Therefore, a linear dependence is the only reasonable description but it may be used only for interpolation. The effect of hydrophobicity of the crosslinker is certainly contributing to this dependence.

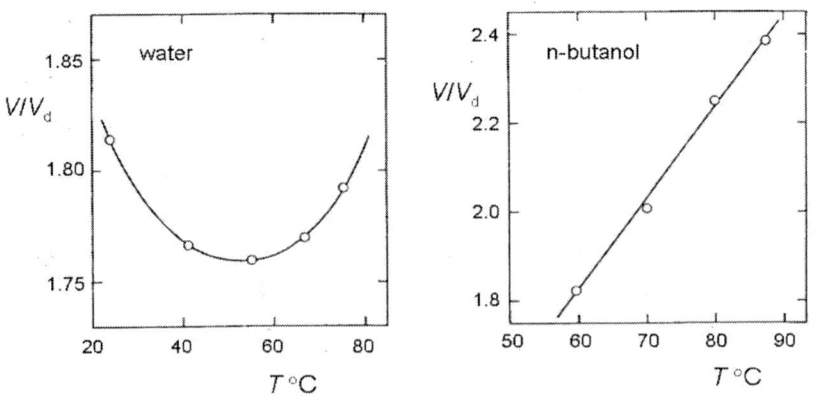

Figure 9. Temperature dependences of swelling of poly(2-hydroxyethl methacrylate) gel crosslinked with 0.2 wt.-% (water) and 0.32 wt.-% (n-butanol) ethylene dimethacrylate and prepared in the presence of 40 wt.-% water. V/Vd - volume degree of swelling. Data of ref. [28].

The dependence of χ for water swollen polyHEMA was found for the range of concetrations of ethylene dimethacrylate 0.1-2.7 wt.-% to satisfy the relation $\chi = 0.32 + 0.90\phi_2$ [29].

Effect of Degree of Crosslinking

It is very well known that the swelling degree decreases with increasing degree of crosslinking of the network. This can be understood inspecting Eq. (3): terms that are positive decrease the degree of swelling (increase ϕ_2). In hydrogels, the crosslinker units (like ethylene dimethacrylate) are hydrophobic and thus also the interaction parameter can increase with increasing crosslinking. To have a feeling of the effect of the degree of crosslinking, some calculated dependences are displayed in Figure 10 and Figure 11. The molar volume is arbitrary: it has been found that for treatment of aqueous systems by the Flory-Huggins theory, the best correlation was obtained if water was considered as trimer or tetramer of H_2O [22].

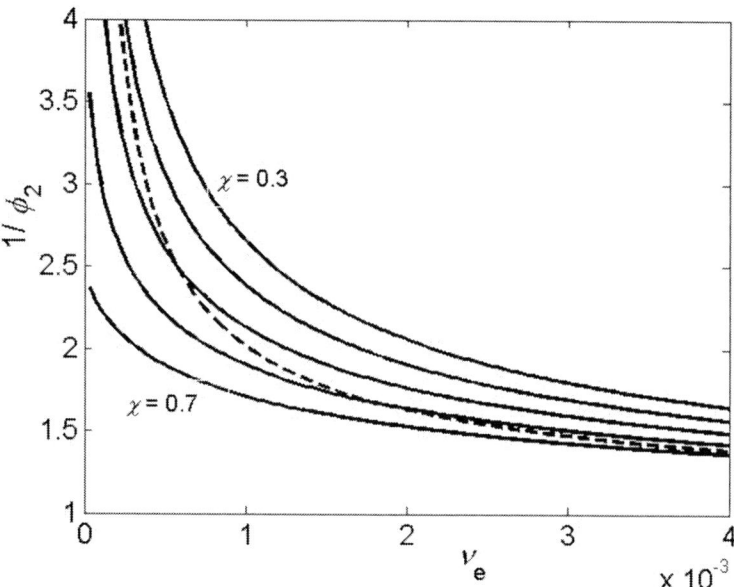

Figrue 10. Dependence of volume degree of swelling according to Eq. (3) on the concentration of EANCs (mol/cm3). Molar volume of solvent V1 = 100 cm3, $\phi 0 = 1$. Full curves are calculated for concentration independent χ = 0.3, 0.4, 0.5, 0.6, 0.7; the dashed curve calculated for the concentration dependence $\chi = 0.4 + 0.7\phi 2$.

A similar dependence illustrates the effect of dilution (Figure 11).

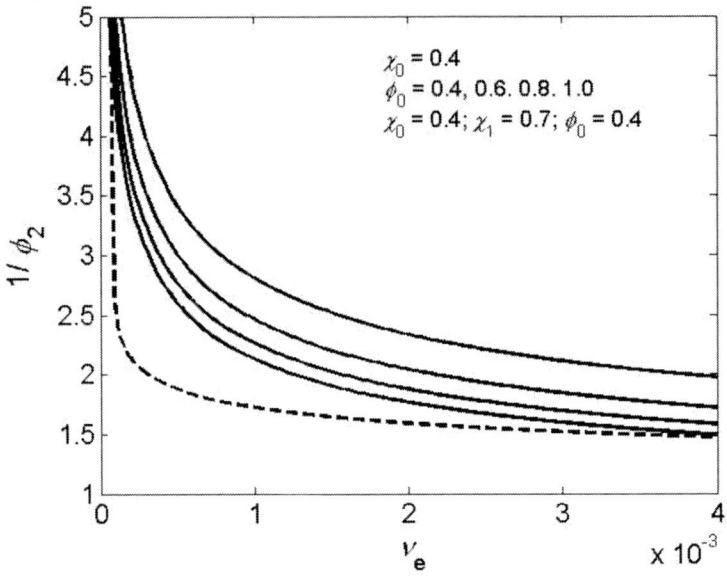

Figure 11. Dependence of volume degree of swelling according to Eq. (3) on the concentration of EANCs (mol/cm^3). Molar volume of solvent $V_1 = 100$ cm^3, $\phi_0 = 1$, 0.8, 0.6, 0.4 from bottom to the top. Full curves are calculated for concentration independent $\chi = 0.4$; the dashed curve calculated for the concentration dependence $\chi = 0.4 + 0.7\phi_2$ and $\phi_0 = 0.4$.

Effect of Dilution During Network Formation

Very important is the effect of *dilution at network formation* characterized by ϕ_0. Progressive dilution increases the degree of swelling. The purely diluting effect is manifest by a change of the chain conformation: the network chains are relaxed at the state of preparation and they become supercoiled in the dry state as a result of shrinkage. Along with the effect on network chains conformation, dilution has some additional effects that tend to increase the degree of swelling: (a) dilution diminishes the interchain interactions and number of entanglements and (b) promotes closing of elastically inactive loops. Attainment of a certain value of ϕ_0 is necessary for the onset of liquid-liquid and liquid-gel phase separation during network formation and facilitates volume phase transition. Increasing *molar volume* of the swelling liquid, V_1, makes the second term in Eq. (3) more positive and the gel swells less. The molar volume of solvent is an important factor particularly in achieving phase separation and manufacture of porous structures. Polymeric additives are much more efficient than their low-molecular-weight analogues.

Swelling in Solvent Vapours

Determination of swelling of gels and hydrogels in solvent vapours is experimentally not so easy as swelling in liquid solvents but it is important scientifically as well as practically. For instance, hydrogel contact lenses during their service can be exposed to a relatively dry environment. The surface is then less swollen and even the transition of the soft gel to glass can set in. This transition will cause a drastic reduction of diffusion rate for the solutes and oxygen. Swelling Equation (3) predicts the effect of reduced vapour pressure - the solvent activity a_1 is equal to relative vapour pressure. Figure 12 shows the calculated effect of reducing the vapour pressure.

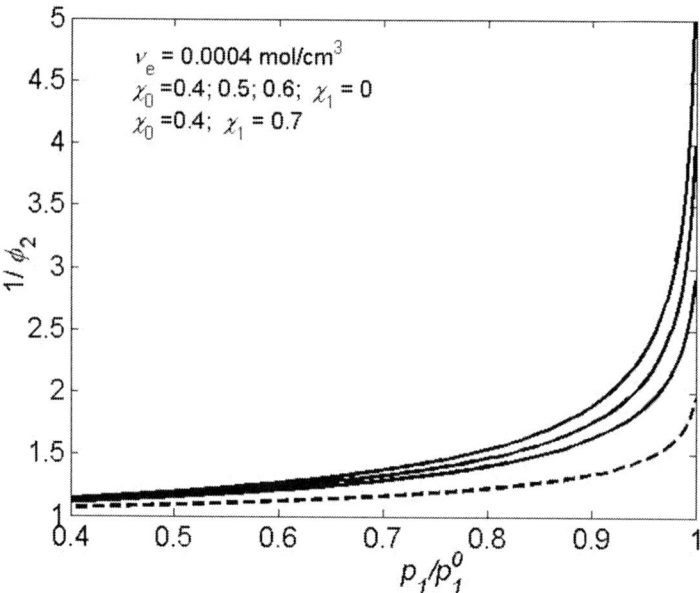

Figure 12. Dependence of the volume degree of swelling of the gel on relative vapor pressure of the solvent; concentration of EANCs 0.0004 mol/cm3, $\phi 0 = 0.6$; different values of concentration independent interaction parameter (full lines) are indicated. Dashed curve corresponds to the concentration dependent interaction parameter $\chi = 0.4 + 0.7\phi 2$.

Characteristic for all these dependences is the sharp increase of the degree of swelling when the vapour pressure approaches saturation. This steep dependence makes the determination of the concentration dependence of the interaction parameter χ difficult. That the gels can approach the glassy state by decreasing the solvent content is another important fact. This rubber-glass transition can cause locking-in of some residual solvent in the glass. This residual amount (several %) is difficult to remove unless the polymer is transferred into the rubbery state by heating.

Swelling in water vapours was studied experimentally using the McBain quartz balances [30],[31] and also in conjunction with glass transition [32]. To avoid problems with water vapour condensation, the measurement was carried out up to $p_1/p_1^0 \approx 0.9$. The measurement confirmed the hyperbolic shape of the theoretical prediction. For polyHEMA, the glass transition temperature of 25 °C was found for about 10 wt.-% water. It can be expected that the main transition region extends to about 20 wt.-% water in polyHEMA. The sorption data were treated using the Flory-Huggins solution theory (Eq. (2)). As expected, the χ-values are high exceeding 1, but they rather decrease with increasing polymer concentration. This can be explained by locked-in excess free volume. Using the Lundberg clustering function, it comes out that in these gels water has a strong tendency to form clusters. When the Flory-Huggins approach is used, clustering is reflected in a strong concentration dependence of the interaction parameter.

Effect of Charged Groups

Presence of charged groups in the polymer gel is an efficient tool for increasing the degree of swelling. Ionized or polyelectrolyte networks carry covalently attached ionized or ionizable groups (e.g., -COOH, -SO$_3$H, --N(Alk)$_2$, -N(Alk)$_3$OH). The degree of swelling of such networks can be very high. Some of them are strong acids or bases and are highly ionized. Some other have to be neutralized to function as fixed ions. Because of condition of electroneutrality, the charge of the ion fixed to the network is counterbalanced by a mobile ion of opposite sign (counter ion) (e.g., -COO$^-$X$^+$, or (Alk)$_3$N$^+$Y$^-$). The ions X$^+$ or Y$^-$ can be exchanged for another ion. Such swollen networks are used as ion exchangers (cation exchanger and anion exchanger) for water treatment and other applications. Hydrogels carrying ionizable groups often exhibit volume phase transition (see below). The reason for a high swelling capacity of polyelectrolyte networks is the hydration of ions, especially of the counterions. In theoretical description of swelling not only mixing of network chain segments with solvent molecules and the elastic response of the network but also the effect of charges must be taken into account. The effect of charges is twofold: the hydration of counterions the concentration of which is controlled by Donnan equilibrium, and repulsive electrostatic interactions of fixed charges that contribute to chain extension. There exist several models to describe equilibrium swelling of polyelectrolyte networks. Usually, the additivity of Gibbs energies is assumed [4],[23]-[34]

$$\Delta G_{sw} = \Delta G_{mix} + \Delta G_{net} + \Delta G_{ion} + \Delta G_{elst} \qquad (7)$$

ΔG_{mix} should include all non-electrostatic interaction, ΔG_{net} should respect the chain extension limit of chains due to high degrees of swelling encountered for polyelectrolyte gels (Langevin function or its expansion instead of Gaussian function). Of two last terms, $\Delta G_{ion} \propto \sum_j (c_j^{gel} - c_j^{ext})$ (the summation extends over all mobile ions) is more important than ΔG_{elst} - the electrostatic repulsion of fixed charges). In equilibrium with external salt solution, the gel phase may also contain co-ions. Addition of co-ion causes screening of the electrostatic field produced by fixed charges and decreases the degree of swelling. To get an impression about the magnitude of the effect of charges the leading term of the polyelectrolyte effect– the Donnan effect - will be added to Eq. (3) and absence of added salt will be assumed. This modification is based on the extended model of Katchalski and Lifson [33],[34]. Several variants of theoretical approach to charged networks can be found in the monograph on the phase volume transition [4]. Using the variant of ref. [33], Eq. (3) is amended by the Donnan term

$$\Delta \mu_1 / RT = \ln a_1 = \ln(1-\phi_2) + \phi_2 + \chi \phi_2^2 + V_1 v_e (A\phi_2^{1/3} \phi_0^{2/3} - B\phi_2) - i\rho\phi_2 V_1 / M_0 \qquad (8)$$

where i is degree of ionization (from 0 to 1), ρ is the density of the polymer, M_0 the molecular weight of unit carrying one charge. This equation is valid for not too large degrees of ionization. Figure 13 demonstrate the effect of degree of ionization on swelling.

The results of Figure 12 show that ionization of only 50 % groups of a network of a 1:1 copolymer of a monomer bearing ionizable groups with an inert monomerand having low crosslink density can increase the degree of swelling ten times.

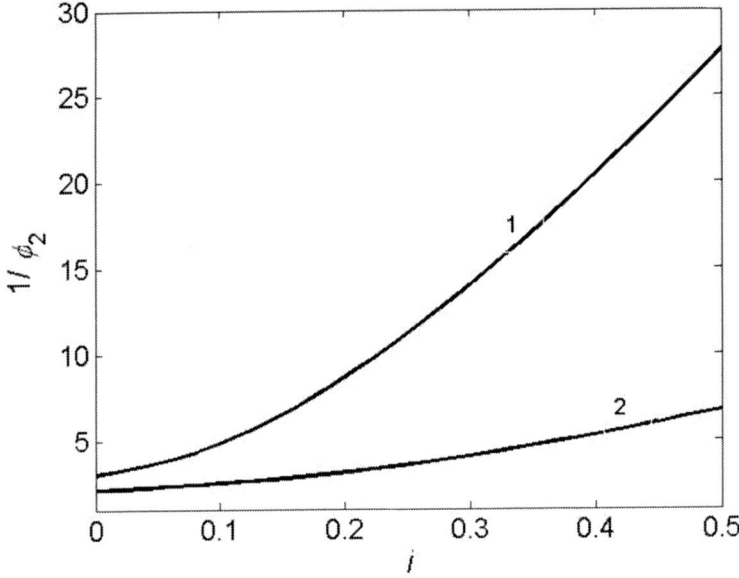

Figure 13. Effect of the degree of ionization, i, on volume degree of swelling; $\chi = 0.5$; $V_1 = 100$, $M_0 = 200$; curve 1 - ve = 0.0005, $\phi_0 = 0.6$; curve 2 - ve = 0.001, $\phi_0 = 1.0$.

However, for such a large expansion one should consider using the finite chain extensibility model as described in refs. [33] or [35].

Swelling Transitions - Two Crosslinked Phases

Analysis of Eqs. (2) and (3) has shown that a situation is possible, at which three phases can coexist: pure solvent in equilibrium with two crosslinked phases of different degree of swelling. By using Gibbs-Duhem relation, the chemical potential of polymer per equivalent segment $m = V_2/V_1$ is obtained in the form

$$\Delta\mu_2/mRT = -\phi_2 + \chi\phi_1^2 + V_1 v_e [A\phi_0^{2/3}(\phi_2^{-2/3}/2 + \phi_2^{1/3} - 3/2) + B(\ln\phi_2 + \phi_1)] \tag{9}$$

In equilibrium, the chemical potentials of components in either phase are equal: $\mu_1' = \mu_1'' = \mu_1'''$ and $\mu_2' = \mu_2''$. Three phases can coexist only in a certain range of parameters of the swelling equation.

For such systems, the calculated dependence of the chemical potential on ϕ_2 exhibits two maxima and the condition for phase equilibrium, equality of chemical potentials of a component in each phase, gave a real solution in the range of volume fractions ϕ_2 <0,1> [36]. Since for simple unionized networks it was difficult to achieve experimentally the proper combination of parameters (sufficiently high v_e at sufficiently high dilution), the experimental discovery of these transitions, characterized by a *jump in the degree of swelling*, was made 10 years later on hydrogels carrying *ionized groups* [37]. Also, it was found experimentally [4] and derived theoretically [25],[26] that nonionized systems with a complex concentration dependence of the interaction parameter (systems exhibiting the "off-zero critical concentration") exhibit this transition. The best known of these systems is poly(n-isopropylacrylamide)-water [38] studied in hundreds of papers.

The volume phase transition can be induced by a number of stimuli, such as change in temperature, degree of ionization (pH), addition of co-ions, change in solvent composition, irradiation, application of electric or magnetic fields. The phenomenon of phase transition is widely utilized in controlled drug delivery. For instance, a gel conjugate when administered *per os* must pass through hostile environment before it reaches its target characterized by a certain value of pH. Then, a sudden expansion in volume exposes the gel interior to agents that split off the drug and rapid delivery of the drug is guaranteed. Other applications include concentration of dilute solutions of higher-molecular-mass, for instance, in technology of pharmaceuticals; for some superabsorbent gels, and various control devices where a fast collapse or expansion of gel-like materials is essential. There exist a number of monographs and reviews on this subject, such as refs. [4] and [39].

Multicomponent Systems

In numerous systems, more components than a crosslinked polymer and one diluent exist. The thermodynamic quality of the mixed solvent can vary linearly with composition, but often passes through a maximum (cosolvency) or minimum (anti-cosolvency). Cosolvency is encountered more frequently. Anti-cosolvency has been found less frequently when the two diluents strongly interact and tend to form a complex. As was already pointed out, all gels formed from a monomer by free-radical polymerization in the presence of an inert diluent are to be considered as ternary or better pseudoternary ones because the monomer has also a function of diluent. The possible situations for ternary systems are shown in Figure 14 using the triangle diagrams.

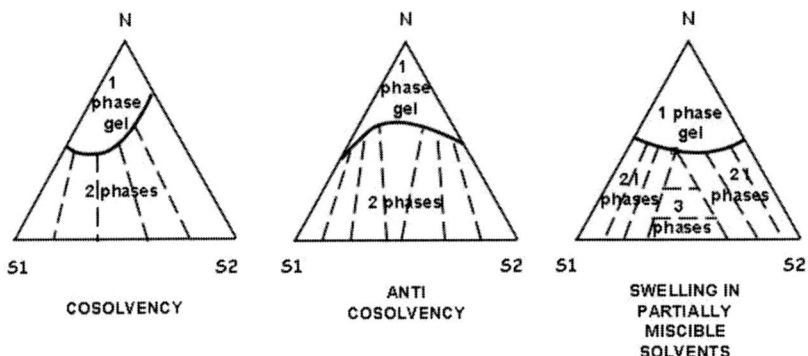

Figure 14. Swelling of a gel in binary solvent; N is network (crosslinked polymer), S1 and S2 are solvent 1 and solvent 2, respectively. The full curve delimits the single–phase region, the dashed lines are tie lines showing composition of coexisting phases.

The diagrams reflect the fact that a crosslinked polymer absorbs only a limited amount of liquids. The first two triangles show systems with maximum and minimum in compositional dependence of the degree of swelling. The third diagram shows the situation when S1 and S2 are only partially miscible and there exists a certain region of coexistence of three phases. One can expect that, for instance, the system crosslinked polyethylene glycol – water – benzene may show up a similar behaviour. More frequent are systems *network polymer – solvent – nonsolvent* with a very asymmetric swelling curve.

Theoretically, the description of multicomponent systems is complicated [25]. The simplest description explained in detail the monograph by Tompa [40] is based on pairwise interactions quantified through three interaction parameters χ_{12}, χ_{13}, χ_{23}. Very often a ternary interaction parameter χ_{123} is necessary, but even then the interaction parameter for mixtures of two liquids may vary with composition. Some other discussion pertinent to polyHEMA-HEMA-water system can be found in ref. [41].

Condition for Phase Separation During Network Formation

In Chapter 2, we have already discussed microsyneresis and macro-syneresis which may take place during formation of a gel.

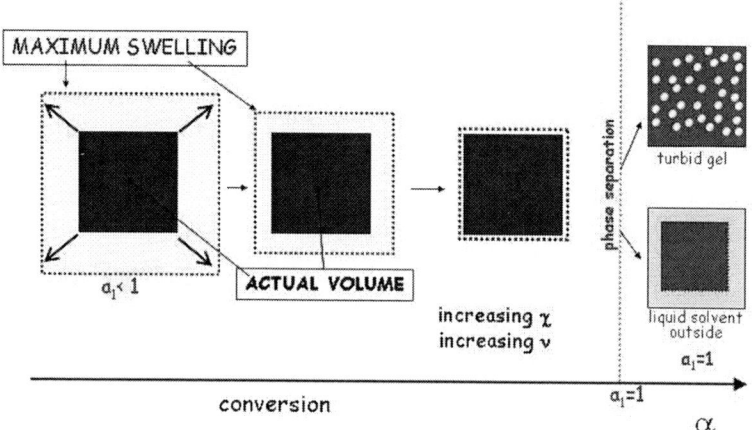

Figure 15. Changes occurring during network formation in the presence of a diluent; a1 is the activity of diluent. For more detailed explanation see text.

It is clear and widely accepted that thermodynamic instability of the system is the reason for phase separation. Intuitively, one can expect that there has to be some relation between the swelling equation and loss of thermodynamic stability because the reacting system is a swollen network in equilibrium with solvent vapor. We will further restrict ourselves to a binary system network polymer – single solvent. We will inspect the swelling Equation (3). The situation can be visualized by scheme in Figure 15.

Dark blue squares show the actual volume of the polymerizing systems (gel swollen not to maximum) which is constant until phase separation sets in; the activity of solvent a_1 < 1. The light gray-blue region shows the hypothetical volume of gel if the reaction were stopped and the gel were brought into contact with excess of solvent. As the reaction conversion increases, the solvent activity in the sample also increases until it reaches the value 1, which corresponds to pure solvent. At this point, phase separation starts. Beyond this point, a_1 remains equal to 1 (it cannot grow over this value) and increasing crosslinking causes separation of more and more diluent. In terms of ϕ_2 and ϕ_0, the incipience of phase separation is defined by [42]

$$\phi_2 = \phi_0 \qquad (10)$$

By inserting the condition (10) into the swelling Equation (3) the critical amount of diluent is defined

$$\ln(1-\phi_0)+\phi_0+\chi\phi_0^2+V_1v_e\phi_0(A-B)=0$$
$$(A-B)_{ph}=\frac{f_e-2}{f_e};\quad (A-B)_{af}=1-\frac{2}{f_e}=\frac{f_e-2}{f_e} \qquad (11)$$

It is interesting to note that the two limits of the Flory-Erman junction-fluctuation theory, one gets the same result – $(f_e - 2)/f_e$ (f_e is the average number of infinite paths issuing from an elastically active branch point), which close to the gel point give the value 1/3, because very close to the gel point f_e = 3 irrespective of the chemical functionality of the crosslink [43].

When applying the condition (11), one has to realize that changes of v_e and possibly χ are the driving force for reaching phase separation. Thus, in applying Eq. (11), v_e is to be taken as a function of conversion and what is actually found is the critical conversion at the given dilution.

3.1.2. Swelling under confined conditions

In many applications, especially in medicine, gels cannot always swell freely because they are geometrically confined by the space available or by developing objects (cells, tissues). The gel objects can be confined in one dimension (swelling between two plates), in two dimensions (gel confined in a cylinder), or three dimensions (gel enclosed in a cavity). The confining wall can be rigid, or elastic yielding somewhat to the swelling pressure. Questions arise: How much does such confined gel swell? How much is swelling anisotropic? If confined, what are the forces generated by the gel and acting against the confining wall?

To predict such effect, we can employ the change of the Gibbs energy (Eq. (2)) defining the deformations along the principal axes x, y, z and find interrelations between the deformation ratios using bounds such as constancy of the dry volume. The anisotropic version of Eq. (3) can serve for calculation of the degree of swelling under constraint. Alternatively, the forces acting on the constraining walls can be calculated by applying the relation between the force, F_i and change in the Helmholtz energy ΔA_{sw} (for atmospheric pressure $\Delta A_{sw} \approx \Delta G_{sw}$)

$$F_i = \frac{\partial \Delta A_{sw}}{\partial L_i} \qquad (12)$$

where L_i is the sample length along the coordinate axis *i*. Confinement during network formation or arising during network formation can affect phase separation and other processes (distribution of chain extensions). In the past, such approach demonstrated the effect of constraint by adhesion on vapor pressure of a solvent evaporating from a coating film during its drying [44].

3.1.3. Other Important Issues

There are several other important issues related to swelling not discussed here. Certainly, it is the dynamics of swelling depending on whether swelling is connected with glass transition or not, and dependences on sample geometry. Also issues of hysteresis in transitions rubbery gel →glassy gel and glassy gel → rubbery gel and the associated excess free volume locked-in in the glass are interesting. Swelling of heterogeneous structures and determination of the swelling degree of the polymer matrix and pore volume in the swollen state are important issues as well. We believe that studies of solvent sorption as a function of vapour pressure can throw more light on this problem.

3.2 Mechanical Properties of Hydrogels

The key mechanical property of hydrogels (and polymer gels) is their equilibrium modulus of elasticity determined in the swollen state. The modulus is a measure of the mechanical resistance of a sample against the external mechanical load. Commonly, in polymer science, the modulus is expressed as a ratio between force acting onto a sample unit area, the stress, and the strain which is the deformation of the sample. The rubbery materials deform under loads reversibly to large deformations that span typically to hundreds of percent of elongation, and their recovery when the loading stops is very fast, instant by the theory. The swollen hydrogels behave to some extent as rubbers, they deform reversibly in a certain range although they rupture usually at much lower deformations compare to weakly crosslinked rubbers. The theory of rubber elasticity establishes the relation between an important structural parameter of the gel macromolecular structure: between the concentration of elastically active network chains and the mechanical response, that is the equilibrium modulus of elasticity. The fundamental thermodynamic concept for this relation was laid in the early days of polymer science by P.J. Flory [19]. The theory of rubber elasticity was being further developed by Mark and Erman [45].

The relationship between stress and deformation of an elastic network reads:

$$\frac{F}{A_{0S}} = G_{sw}(\lambda - \lambda^{-2}) \qquad (13)$$

where F is force applied on a sample measured for instance by a tensile instrument, A_{0S} is the cross-sectional area of a non-deformed swollen sample (before the measurement), G_{sw} is the equilibrium shear modulus and λ is the deformation expressed as L/L_0. L_0 is initial length (before deformation) and L is length of the deformed sample.

From the value of G_{sw}, obtained from the stress-strain measurement by Eq. (13), the so called crosslink density (v_e) can be calculated using a theoretical equation for a swollen network in equilibrium:

$$v_e = G_{sw}/[RTA_f(\varphi_2^0)^{2/3}(\varphi_2^{1/3})] \qquad (14)$$

where R is universal gas constant, T is absolute temperature (K), A_f is a dimensionless constant that characterizes the functionality of a crosslink. The φ_2^0 term is a volume fraction of macromolecular network at the network formation ($\varphi_1^0 = 1 - \varphi_2^0$, where φ_1^0 is the volume fraction of diluent present in gelling system). The φ_2^0 term called also "memory term" characterizes a state of macromolecular chains at the condition of network formation and is particularly significant for hydrogels as they are often prepared in diluted systems. The variable φ_2 is the volume fraction of macromolecular network in the swollen system macromolecular network–solvent in equilibrium and must be known for the given system for the same swelling solvent as at which the mechanical test is performed. The obtain v_e in common units moles per cm^3 from the Eq. (14), one can for instance use the units: megapascals for G_{sw} and a factor of 10^6 in the denominator.

Instead of the not exact term "crosslink density" rather a well defined expression "concentration of elastically active network chains, EANC" should be used and understood. The language here grasps the essence of the thermodynamic relation; the equilibrium modulus in rubbery state is proportional to the number of molecular connections between the crosslinks per unit volume. The Eq. (14) gives the number of EANC per unit volume of a dry network which is a rational reference state at which the comparison of various systems should make sense. It is possible to derive from the same assumptions a relation for the number of EANC per volume of the swollen network (only the exponents will differ).

The constant A_f, so called front factor, falls into a range <0.5 - 1> and it quantifies the effect of the average functionality of crosslinks on the equilibrium deformation behaviour. In the so called affine network model, where the macroscopic deformation represents a sum of local microscopic deformations, the A_f value is equal to 1. In the so called phantom type model network the crosslinks are considered as fluctuating volume elements on a certain length scale, the value of A_f is defined as $(f_{ave}-2)/f_{ave}$ where f_{ave} is an average crosslink functionality. Because the minimum number of network chains extending from a crosslink is three to form a three-dimensional infinite network, the lower limit of the range for A_f is 0.5. To decide on which type of network should be considered to interpret given data, either affine of phantom, is not always a quick task and it may demand additional experiments or literature search on the properties of relevant systems. There is a vast literature available containing measured characteristics for many types of polymer networks and gels. In practise, the limit A_f value of 1 will serve well for an evaluation of a majority of covalently crosslinked swollen hydrogel samples. The comprehensive explanation of the theory of rubber elasticity can be found in the excellent books by Mark and Erman [21],[45].

Sometimes a misleading consideration about the molecular weight of network chains between crosslinks appears in the literature and textbooks on polymer science. An oversimplified relation $M_c = RT\rho/G$ is sometimes applied where M_c is meant to represent the average molecular weight of chains between crosslinks. This relation would be valid only for an ideal (dry, not swollen) network where all the chains are perfect and no dangling chains or inactive loops and cycles appear. In real networks, the connection between crosslinks can have various lengths, branching, or they can bear dangling chains of different sizes on them. There can be unanchored chains pointing out from a crosslink. The expression in such instances is of course not valid. Therefore the structural characteristic parameter considered from equilibrium modulus measurement should always be the number of EANCs in unit volume (of dry network).

It should be noted that to get correct values of the equilibrium modulus, the measurement should not exceed the so called linear viscoelastic region that is a region where the relation between stress and strain is linear.

The equilibrium modulus can be measured either as a strain responding to introduced stress or vice versa, as a stress responding to defined strain. The equilibrium, i.e. the time independent values of the responses should be always sought for obtaining the structural parameters of the macromolecular network correctly. Relaxation or creep measurements are good choices.

There are several possible ways of the geometrical arrangement of the modulus measurement. It can be measured in tensile mode by applying force in the direction of the long dimension of a bar-shaped sample (Figure 16). The practical advantage of this way is the possibility of higher force response at very low deformation (high sensitivity) and well defined geometry of a sample. The sample held in clamps can also be loaded in torsion that means that usually one clamp rotates and thus deforms the sample. However clamping a soft hydrogel may be the bottle neck of the method even if a testing device of sufficiently high sensitivity (for soft matter) is found. The swollen sample can easily be drawn out from the clamps or can burst in the clamped part. The very soft samples will sag.

Such technical issues may be partially solved by use of torsion test geometry (Figure 17). An environmental chamber with temperature control and solvent excess into which the measuring geometry with the fixed sample is immersed during the test should be always applied.

There have been successful apparatuses contrived during the golden age of polymer physical science in the third quarter of twentieth century, in the predigital era. For example Cluff et al [46] constructed a simple compression modulus measuring device using a micrometer gauge and a series of weights. The working part of the apparatus, the two parallel plates of a plunger with the fixed pellet sample, could be immersed in a solvent during the measurement to prevent evaporation. The advantage of this design was its simplicity and a high precision at the same time and yet the device is quite inexpensive. In many laboratories who deal with hydrogel characterization such handy apparatus is used even today and its use is fully justified.

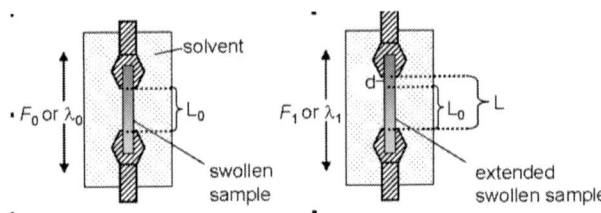

Figure 16. Tensile arrangement of the stress-strain measurement of a swollen hydrogel sample.

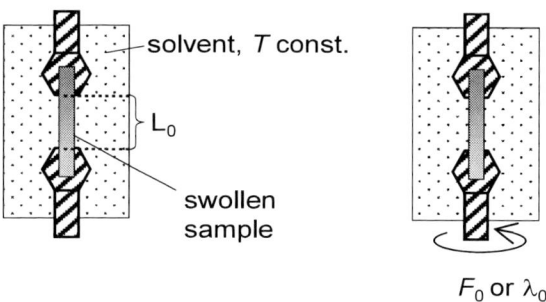

Figure 17. Stress-strain measurement in torsion. The torsion angle is very small, typically lower than one degree.

More sophisticated testing machines appeared in the early nineties. Their output became digital and the devices such as tensile modulus machine or various types of rheometers became computer-instrumented.

Nowadays, the controlling software and the hardware of fine multiple task machines for viscoelasticity characterization are inseparable elements. The common laboratory device suitable for hydrogel characterization is an oscillatory rheometer. Various measuring geometries are available including solvent traps or humidity control chambers especially suitable for testing water swollen gels. The basic measuring mode is shear with the control of the force in normal direction and precise gap measuring and setting Figure 18. The samples have then disk-like or cylindrical geometry. The majority of contemporary rheometers is equipped with the possibility to monitor the steady strain or stress responses in time, i.e. to perform the relaxation measurement leading to equilibrium modulus value and they are capable of shifting between strain controlled and stress controlled modes. To find the equilibrium modulus, the sample is in contact with the upper measuring plate. By turning only the upper plate around its central axes of a small angle, a small deformation is introduced on the sample. The rheometer reads the sample force response - the value of stress exerted by the sample when the plate is kept steadily in its new position. Under the load, the chains in the hydrogel will rearrange to spend minimum energy for holding in the new position. This process is called strain relaxation. The value of sample stress response that will not change in time is the value needed for finding the equilibrium modulus. A similar experiment can be done in a creep mode. The plate is pushed to turn around the central axis with a constant force set before. The sample response is a certain deformation. When the final deformation is reached, the so called creep compliance or creep modulus can be calculated.

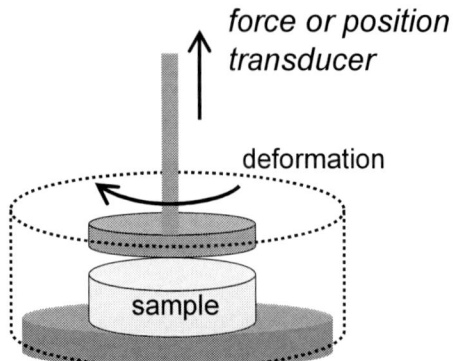

Figure 18. The scheme of shear oscillatory rheometer measuring geometry.

However, the basic mode in rheometry is dynamic, from which the value of moduli is not equilibrium but frequency dependent. Sometimes the frequency influences the modulus only negligibly so when the value is very close to the equilibrium modulus, the result from experiment is the "quaziequilibrium modulus". When viscoelastic materials (as most of the polymers are) get loaded in a dynamic manner, typically with sinusoidal oscillations at set angular frequency ω, a very complete information about their structure can be figured out from the mechanical responses. The whole scientific field of mechanical spectroscopy of polymers has deveoped.

The sample response is divided into a real (in-phase, storage) contribution, and into an imaginary (out-of-phase, loss) contribution. The real part gives so called storage modulus, G', which is controlled by elastic properties of material; and out-of-phase (loss) part, giving loss modulus G'', which is a measure of viscous properties of material. For instance by application of the oscillatory strain $\gamma(t) = \gamma_0 \sin(\omega t)$ it is possible simultaneously and independently obtain the both moduli:

$$\sigma(t) = \gamma_0 (G'(\omega)\sin(\omega) + G''(\omega)\cos(\omega t)) \qquad (15)$$

The G' and G'' moduli are related according to equation:

$$G^* = G' + iG'' \qquad (16)$$

In the oscillatory rheometer, a viscoelastic measurement of swollen hydrogels and polymers is usually performed in the parallel plate-plate or sometimes cone-plate geometries. Dependences of the oscillatory strain $\gamma(\omega)$ or oscillatory stress $\sigma(\omega)$ are typically plotted together with the phase angle $\delta(\omega)$ that characterizes the shift between the real and imaginary parts:

$$G = G^*\cos(\delta); \quad G' = G^*\sin(\delta);$$
$$\tan\delta = \frac{G'}{G} \tag{17}$$

For the rubbery materials with a fully elastic response, the phase angle δ will be 0° and for the purely viscous liquid (with no elasticity) the phase angle will be 90°.

The dynamic oscillatory mechanical spectroscopy is a powerful tool for characterizing the time changes in materials; very importantly for hydrogels: their structure build-up during crosslinking polymerization. A major contribution to this topic was the work of Chambon and Winter [47] published in 1985. In this paper, they first discovered that in the vicinity of the gel point (called rather a critical point), the time dependences of $G'(\omega)$ and $G''(\omega)$ plotted logarithmically were parallel. They presented convincing experimental results obtained with model polybutadiene polymer networks. The gel time thus can be well characterized using the multiple frequency rheometry.

An excellent introduction to viscoelasticity of polymers can be found in the comprehensive classical textbook by Ferry [48]. The details on rheometry, the painstaking description of experimental equipment technique with the theoretical background is available in the book by Macosko [49]. The issue of rheological characterization of polymer networks has been recently very clearly worked out with many important references in the textbook by Pascault at al. [50].

One practical draw back of the rheological characterization of a material is the relatively high volume of material needed for testing. Importantly, the rheology will not give quantitative results with porous hydrogels it only describes the sample as a whole without taking into account its internal microstructure.

A curious and a promising method for hydrogel characterization is the so called microrheology. Microrheological methods have appeared in the year 1948, when the Einstein's theory of Brownian motion was proved experimentally by Jean Baptist Perrin and when the mean-square-displacement

(MSD) of microsized beads in viscous liquid was observed and measured using microscope for the first time. The historical background of the early experimental work as well as the recent development of the theory and techniques are comprehensively described in the review [51]. The microrheological methods experience their comeback in the present when much more sophisticated in-situ observation technique is available. The principal of microrheological measurement is simple; nowadays not one particle but rather an assembly of small particle-probes embedded in the tested material is being traced and the mean time displacement of particles is calculated from the time dependent image analysis and plotted as a time function. Usually as probes submicron sized latex particles that can undergo the Brownian motion (i.e. submicron size) are used. In a viscous liquid, the particles would travel apart each other in time without limitations, so the mean-square-displacement will steadily increase. With the growth of liquid viscosity (e.g. with gelation), the probes would become more and more fixed by the structure in a certain volume and finally, as the system goes through its gel point, the characteristic displacements of the probes will become constant. In the active microrheology, the particle movement in the sample is excited by the external force, such as the magnetic field. Knowing the external force extent and the displacement from measurement, the moduli of the sample can be calculated. The microrheology brings advantage mainly for biological materials science when often only small amounts of samples are available and can serve mainly as a semi-quantitative characterization of the gelation process. There is a well developed web side about the microrheology authored by the scientists of the Department of Physics of the Harvard University [52]. Theoretical fundamentals of microrheological methods are further reported in [53],[54].

The absolute values of moduli of the synthetic hydrogels in their fully reacted swollen state are very important factors informing about the possibility of *in vivo* gel use. Also, when cells are supposed to attach and to spread onto the artificial gel support; the mechanical properties of such support seem to be the key factor. It has been evidenced that the living cells feel and prefer the stiffness of the environment in which they proliferate [55]. The literature brings numerous examples of experimentally determined mechanical moduli of various tissues; typically the gel stiffness is described with compressive or shear moduli. Helmlinger et al. [56] studied brain tumor proliferation and proved that multicellular tumor spheroids can overcome the mechanical stress up to 6 kPa in an agarose gel before they become inhibited at stresses between 6 and 16 kPa. Yu and Schoichet [57] studied hydrogel scaffolds (prepared from a porous synthetic hydrogel) for the growth of neurites from primary neurons in order to regenerate soft nervous tissue. They state values of compression modulus of such tissue to be around 200 kPa.

A homogeneous slightly crosslinked hydrogel based on the poly(2-hydroxyethyl methacrylate) prepared with 30 wt.-% of water in the reaction mixture as diluent had the relaxation (equilibrium modulus) measured in shear around 55 kPa. When the same hydrogel is made highly porous by increasing the dilution its total modulus decreases down to 2.5 kPa.

This part of the chapter could only serve as a brief introduction into the issue of hydrogel mechanical behaviour and its characterization. To gain further knowledge, the reader might appreciate the very good review by Anseth [58] with its some fifty collected references.

Chapter 4

Contact Lenses

Contact lens is a small optical device placed directly on the cornea. The technology of its production must allow tuning of optical properties, it must be possible to adjust shape precisely and ensure appropriate physiological function of lens in contact with the eye. The cornea is a transparent tissue with no veins that receives the nourishment mostly from the air oxygen. The artificial contact lens, however, always has some effect on the regular metabolism of the cornea. It is necessary to minimize the hypoxic and mechanical stress that may occur upon contact lens use.

The first idea to alter corneal power by optical system placed directly on the cornea was described by Leonardo da Vinci in the beginning of sixteen century [59]. His solution consisted in immersing the eye in a bowl of water. Next generation of similar constructs has been described by René Descartes in 1636 [60]. Twinkling of eye represented however an unsolved problem. In 1801 Thomas Young [61] constructed fluid-filled tube equipped with a glass lens, which was placed to the eye orbital rim. Sir John Herschel in 1845 proposed glass contact lens analogous to modern lenses, but the space between cornea and the lens was sandwiched with animal jelly [62]. More than forty years after this, the first real contact lenses of glass were made. The progress made of in the area was associated with the names of Adolf Eugene Fick (Switzerland), Eugene Kalt (France) and August Müller (Germany) in the late 1880's. The half walnut shaped lenses were made by glass blowing, they were scleral and afocal and they rather served as a partition between cornea and eye lid to prevent spreading of a disease mechanically. Their preparation, testing and application were described in detail by Adolf Eugene Fick in 1888 [63]. In the same year, a successful use of such lenses on human patients suffering from keratectomy was reported in France by Eugene Kalt [64]. First sanded glass contact lenses had defined optics and could correct the eye refractive error. Their inventor, August Müller used them to correct his myopia and he published his experience in 1889 [65].

Another significant progress in the contact lens development was made in 1936 when a new plastic, polymethyl methacrylate (PMMA), was brought to the market by Rohm and Haas company. In this year, Wiliam Feinbloom [66] reported on a scleral lens made from plastic and glass parts. Soon after, a first fully plastic lens was manufactured using lathe-cutting and polishing techniques. This was the beginning of the new whole area of industry. The transparent synthetic polymers became the suitable class of materials for contact lens production.

During the World War II the PMMA was shown to be biologically inert and it was also less fragile compare to glass. These reasons together with easy processability caused that PMMA became a key material for contact lens production for several decades. Lathe-cutting allowed making lenses of any shape and optical power.

In 1950 Kevin Tuohy patented the hard corneal PMMA lens [67] used in its somewhat advanced version to treat certain conditions even today. They are known under the term rigid gas-permeable lens (RGP).

Over all the advantages, one drawback of PMMA plastic was that the material was not permeable for water neither for gases. That means that no oxygen or any water-soluble compounds could not penetrate through however the cornea needs to be doped by the air-oxygen permanently. If for longer time, the cornea suffers from hypoxic stress, the malfunctioning and serious damage appear. In the course of solving such problem of the material, the siloxane structures or perfluorated alkylmethacrylates within the PMMA chains were investigated [68].

Silicone elastomers are known for their high gas permeability including oxygen. Also, these soft materials have some mechanical parameters comparable to natural tissues. This was an important feature as it became known that the mechanical stress can be as serious as the hypoxic stress. On the other hand, the soft material could not be shaped by the lathe cutting; the mold pressing into the precise individual molds had to be used instead. Another drawback of silicone elastomers is their hydrophobicity. Water-soluble species will not penetrate easily through a silicone layer while the adhesion to the eye surface is very strong. Some patients even experienced severe damage of their cornea when removing silicone lenses from their eye. The highest number of patents of silicone contact lenses was filed around 1965 and the development went through its peak period in late seventies. But after all, for the disadvantages mentioned above, the silicone lens did not make its turn. Yet, it is still used but only sparingly; in cases that demand a special treatment.

The rigid gas-permeable lenses (RGP lenses, PMMA-based) are still favored and commonly used in many developed countries. The development of gas-permeable plastics still continues and the values of permeability for oxygen of the current RGP lenses are quite high while the surface wettability is adopted precisely to application. They are often manufactured to meet individual needs, i.e. tailored to each patient. They are still indispensable in the correction of high astigmatism and for clients with keratitis. Their development is historically associated with a number of patents by Norman Gaylord [69].

A remarkable achievement in the lens technology was the discovery of soft synthetic hydrophilic polymers, synthetic hydrogels, that were used for first lenses in their lightly crosslinked state [70],[71]. The polymeric synthetic hydrogels have the capability to absorb water up to certain equilibrium volume that is constant for given system and temperature (equilibrium swelling). Therefore, the hydrogel contact lens contains always some water that functions as a transport media for low molecular weight nutrient species and metabolites (oxygen, ions, etc.). Even though the number of users of the second and third generations of the silicone-hydrogel contact lenses is increasing remarkably and their market is undergoing steady growth, the (methacrylate)hydrogel lenses are nowadays the most common type of contact lenses. This is perhaps for the flexibility of their application regimes.

The polymer material for hydrogel contact lenses were developed in fifties by Otto Wichterle and Drahoslav Lim [70]. Professor Wichterle patented a basic material, poly(2-hydroxyethyl methacrylate), polyHEMA for production of hydrogel contact lenses together with a unique method of their manufacture; the spin-casting [71]. In 1963 Wichterle added a patent for lathe-cutting of soft contact lenses from a hard block of dried gel; xerogel. Although the first applicable contact lenses were spin-casted in 1961, the production was started in the USA in 1972 by the Bausch and Lomb Company.

Hydrogel materials integrate several important properties: they are soft, wettable, they are compatible with living tissue, they are partially permeable for gases, and they allow flow of water through their structure as well as diffusion of low molecular weight solutes. They also have some drawbacks. The original first hydrogel material, crosslinked polyHEMA, can swell in water at the laboratory temperature only up to about 40 wt.-% of water content in the gel-water system. Its permeability for oxygen is about 8-12 barrer (whereas the RGP material permeability can be 8-120 barrer and that of the silicone elastomers up to 200 barrer) [72].

Since their discovery, the hydrogels were further investigated with the goal of the equilibrium water content and permeability for oxygen increase. Thus, the copolymers of HEMA with other hydroxyalkyl methacrylates, methacrylic acid or its salts were prepared, cf. Figure 19. These hydrogels swelled in water up to 55-60 wt.-% of equilibrium water content while their Dk reached values in the range 20-25 barrer, nevertheless they were prone to attachment of protein deposites. The protein adhesion is promoted by the lower water content in the material surfaces and is stronger when there is a charge in the macromolecular chain of the substrate (cf., e.g. methacrylic acid and their salts). The higher water content reached materials based on the copolymers of N-vinylpyrrolidone with different alkyl methacrylates, or hydroxyalkyl methacrylates. These polymers can attach lipophilic deposites that are however more easily removable. The group of hydrophilic monomers for contact lenses includes many other molecules, e.g. glycerol methacrylate, glycidyl methacrylate, (meth)acrylamide, poly(vinyl alcohol).

The development of materials for contact lenses spanned over three decades. Within that time not only the chemistry but also the medical design of the lens body and the wear regime would be subjected to changes.

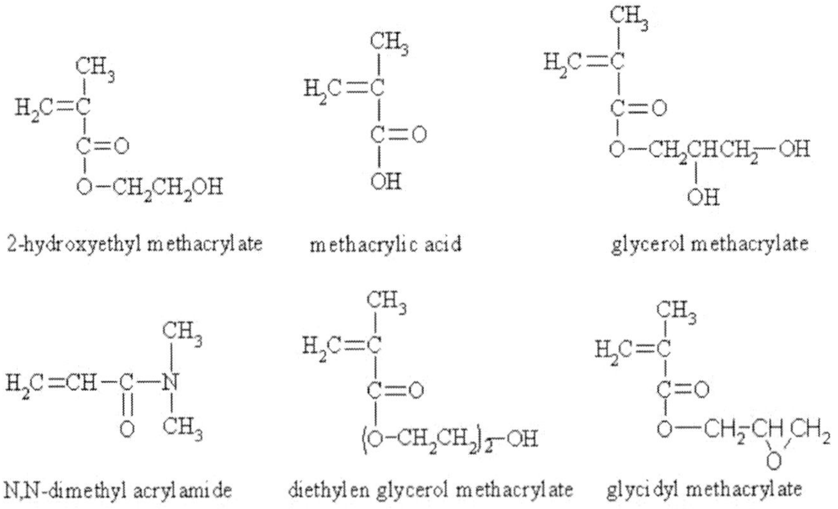

Figure 19. The chemical structures of methacrylate monomers.

In 1998, the first lenses with planned replacement were introduced to the market, Acuvue (J and J) and six years later, 1994, the first daily lenses were brought up (Bausch and Lomb) [73].

a)

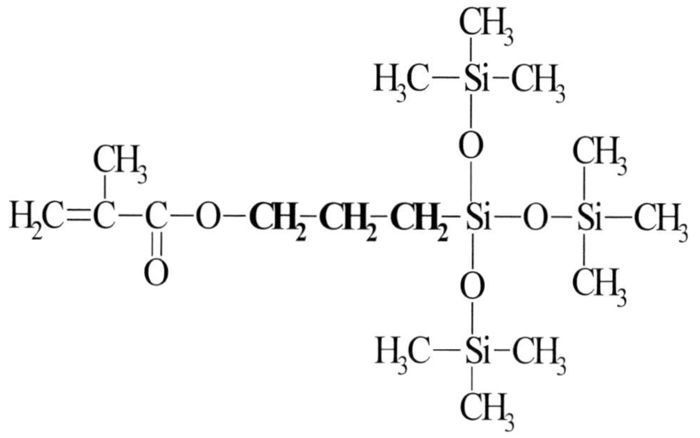

b)

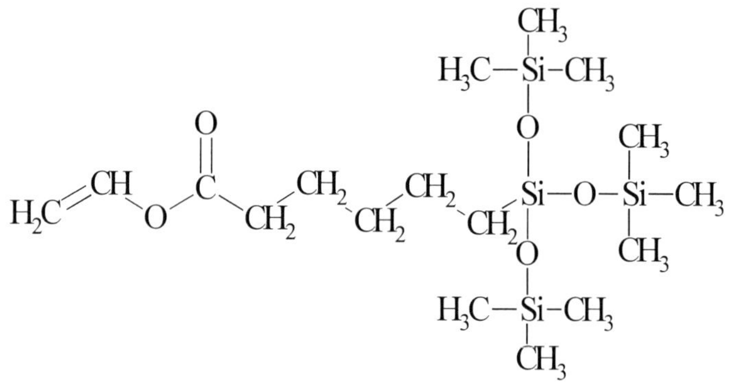

c)

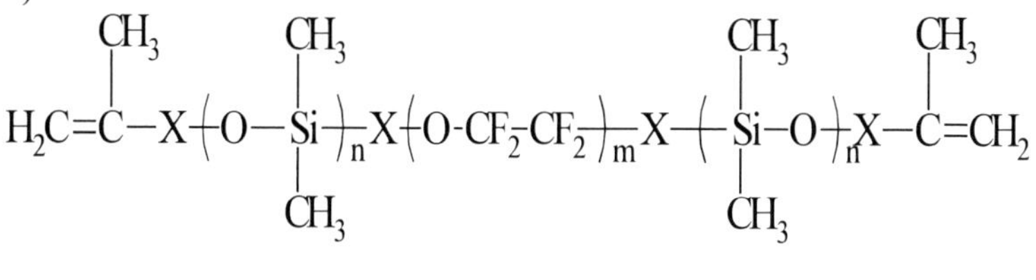

(n = 5 -100, with advantage 14 - 28; m = 10 - 30)

Figure 20. The examples of chemical structures of monomers for silicone hydrogels.

Since the beginning of seventies, the contact lens material design was driven by the idea to combine the advantageous properties of the silicone elastomers with the hydrophilicity and thus have a transparent, highly permeable, and swellable material with the capability to allow diffusion of small species. Various sandwiched materials were constructed or surface modifications were tested, the most common approach to that was hydrophilization of the silicones. Most of the attempts turned unsuccessful. The break through was seen at the end of the nineties. The research teams around the world would focus on making a co-continuous structure, silicone hydrogels [74].

The high oxygen permeability is achieved with the monomer commonly referred to as "TRIS" (Figure 20A), which is known from RGP materials. The methylene groups in the structure of TRIS (in bold) represent the sites for hydrophilic modification. Material balafilcon A (PureVision CL) is based on polymer of vinyl carbamate derivative of TRIS (Figure 20B). Lotrafilcon A (Focus Night and Day) is copolymer of TRIS monomer with N,N dimethyl acrylamide and macromonomer B (Figure 20C) [74].

Second type of novel silicone-based materials contains oligomers bearing short blocks of hydrophilic chains, oligosiloxanes, and/or perfluorated chains in their backbones. Not only the copolymers themselves were subjects of the patents but also the methods of their processing, e.g. the methods necessary surface modifications to ensure uniform wettability of lens surface, i.e. methods leading to elimination of hydrophobic domains.

In 1998, the first brands of silicone hydrogel contact lenses entered the world market: Focus Night and Day by CibaVision Company and PureVision by Bausch and Lomb. Application of these materials in the end-use products has set another milestone after the invention of the synthetic hydrogel contact lens by Wichterle. The advanced silicone hydrogels achieve swelling values comparable to standard hydrogels, their permeability for oxygen is lower but their surface hydrophilicity is enhanced (wettability), their Young modulus is lower - these are silicone hydrogels of the 2nd and 3rd generation.

The current trends in the field of standard and silicone hydrogels focus on inclusion of a highly hydrophilic linear monomer bound into their molecular structure by physical interactions instead of covalent bonds. For example, such interactions can have even topological cause - they can be molecular entanglements that prevent the molecular chains from diffusing out of the structure (e.g. by washing) because of their high molecular weight and sterical hindrances. The hydrophilic chains bind firmly water molecules and thus prevent the whole system from drying out while worn.

The contact lenses are modern, sophisticated, and when properly used, a safe vision correction aids. Although a number of materials were described in this chapter, the standard synthetic hydrogels still represent the most typical material for soft contact lenses while the contact lens has been a long-term successful most spread application of hydrogels in biomedical area.

Chapter 5

Intraocular Lenses

Intraocular lenses (IOL) are implanted into the eye in order to correct vision. Their application represents one way of refraction correction or they are used in the cases of aphakic eye resulting from cataract surgery. Cataract operation consists of several phases. A cloudy natural lens is removed and subsequently replaced by a synthetic lens. The IOL parameters have to be precisely adjusted according to the previous measurements of patient's eye [75],[76].

Intraocular lenses can be implanted into the various parts of the eye, e.g. into the interior chamber, the posterior chamber, or the stroma. The implantation locus determines the lens shape parameters. The two main designs of IOLs in current use are the interior chamber IOL and the posterior chamber IOL [76]. Every IOL consists of two parts - an optical part and haptics anchoring the lens. Anterior chamber IOLs lie in front of the iris and have a flexible or semiflexible angle-supported haptics. Implantation of these IOLs is possible for both intracapsular and extracapsular cataract extraction. They are useful as a standby if the posterior capsule was accidentally ruptured (i.e. posterior chamber implantation is not possible). Posterior chamber IOLs lie behind the iris and have flexible haptics which are inserted either into the capsular bag or into the ciliary sulcus [77].

Material for IOLs should fulfill various requirements such as the proper mechanical properties, the refractive index, the glass transition temperature and the biocompatibility. Two groups of materials are used for manufacturing of IOLs – silicones and polymers of acrylic acids. The group of acrylate polymers can be divided into the rigid (hard) materials, such as polymethyl methacrylate (PMMA), and the soft/foldable materials, including hydrophobic and hydrophilic acrylics. The main advantage of soft materials is the possibility of their deformation by folding and subsequent implantation through a very small incision. During the implantation, the lens is placed into the capsule, then relaxes and reaches the previous shape. With haptics, it is fixed in the optical axis of the eye. During the IOL development, the production volume of soft materials dramatically increased compare to hard acrylics due to the possibility to implant soft lenses through significantly smaller incisions with the advantage of lower number of the post-operation complications. However, small or even tunnel incision is more difficult to manage and is more dependent on the instrumentation; that means that this way is more expensive and therefore more common in the rich and industrialized countries. So, the population of the "third world" countries maintains the number of implanted conventional PMMA IOLs still at a high level. Besides the foldable hydrophobic acrylates, e.g. Acrysof® (Alcon Laboratories, Fort Worth, Texas, USA) or Sensar® (Advanced Medical Optics, Santa Ana, California, USA), with the content of water lower than 1 wt.-%, hydrophilic copolymers containing 18-38 wt.-% of water are used [78],[79], e.g. MemoryLens® (Ciba Vision, Duluth, Georgia, USA), Centerflex® (Rayner Intraocular Lenses, Brighton-Hove, East Sussex, UK) and Hydroview® (Bausch and Lomb, Rochester, New York, USA). Hydrogel IOLs are prepared for example from slightly crosslinked network formed by the terpolymer: 2-hydroxyethyl methacrylate-*co*-ethylene dimethacrylate-*co*-methacrylic acid transferred into sodium form [80].

In the case of hydrogel IOLs, the deformation prior to implantation may be fixed and intensified by partial drying and/or cooling. After the implantation, partially dried hydrogel lens increases volume due to the swelling in eye liquids and/or increasing temperature (35-36 °C). Another advantage of acrylic materials is higher refractive index compared to silicones which enables thinner construction of the lenses yet maintaining good refractive properties [79]. The surface of hydrogel IOLs can be modified in order to reach better biocompatibility and to avoid inflammatory reactions after the implantation [81],[82]. The retina can be protected by chromophores absorbing UV light, such as benzotriazole and benzophenone, added to the IOL material [79]. Hydrophilic acrylic material used for preparation of IOLs can also serve as a drug delivery system, e.g. as carriers of antibiotics preventing post-operative complications [83].

The response of the surrounding tissue to the IOL implantation and consequent complications were extensively studied [84],[85]. The influence of hydrophilic and hydrophobic materials, as well as IOL shape, on the creation of secondary cataract as the most common post-operative complication was observed [81],[84]. These studies emphasize the importance of shape, while other source [86] referred that hydrogel IOLs show five times lower occurrence of the secondary cataract in comparison with PMMA IOLs. Also, the reaction of the IOL hydrogel material on the surrounding tissue was investigated [87] and processes such as opacification [88] and calcification [89] were interpreted. Presence of carboxyl groups (especially on the material surface) plays a positive role in decreasing of calcification probability [90].

Development of IOLs has proceeded from relatively simple PMMA implants via flexible materials implantable through a minimum incision to the lenses with more sophisticated optics, e.g. toric IOL correcting astigmatic vision (AcrySof® Toric, Alcon Laboratories, Fort Worth, Texas, USA).

The next important achievements in IOL development are accommodating IOLs, which could provide excellent vision at all distances (far, intermediate, and near) without any side optical effects such as unwanted retinal images, halos, glare, and loss of contrast sensitivity. Moreover, they have the potential to reduce the patient dependence on glasses after cataract surgery [91].

As other advancement, it is necessary to mention the development of light-adjustable IOLs, which affords the opportunity to correct post-operative refractive errors [92]. After IOL implantation, the use of light exposure may initiate polymerization of remaining unpolymerized subunits in the lens material or additional material crosslinking in a precisely defined area resulting in a change in the overall optical lens power.

The future in IOL, as in other areas of synthetic implant applications, is associated with self-assembling systems, it means with the liquid injection systems, which can form pre-defined structure in the desired location. However, cataract operation nowadays represents ordinary matter, which concerns a wide range of populations, particularly in the context of increasing number of refractive surgical interventions. The manner of its implementation, including the selection of implantable lens material, is significantly influenced by the economic possibilities of the health system, respectively patient.

Chapter 6

Functional Implants

Hydrogels for Urinary Incontinence Treatment

Urinary incontinence causes serious medical and social problems, having a deep influence on a patient's psychological state and causing a series of direct social impacts. The management of incontinence is very expensive; the cost in USA in 2000 exceeded 30 billion USD [93]. All current surgical methods of incontinence treatment are based on the increase of urethral resistance at the site of injured or missing sphincter in males and in females with intrinsic stress incontinence. Drugs can not effectively modify the pressure of an injured sphincter in cases of serious muscle damage. The approaches to the surgical procedures in such patients are based on the increase of urethral resistance in the sub-vesicle area by the application of teflon, collagen, or silicone particles [94]-[96], the sling [97], the artificial sphincter [98] or the urinary catherization [99]. Such artificial obstructions are aimed to prevent spontaneous leak of urine, whenever the abdominal pressure increases; on the other hand, it has to enable free passing of urine without residuum.

The use of hydrogel implants in incontinence treatment brings very promising results [100],[101]. The method consists in subcutaneous implantation of a cylindrical dry hydrogels (Figure 21a) to the damaged area of sphincteric part in bulbomembranous urethra. The number of implants is chosen depending on the extent of incontinence and the swelling degree of implant is adjusted. Dry implants get swollen by extracellular liquid and thus create the obstruction, which prevents accidental escape of urine (Figure 21b).

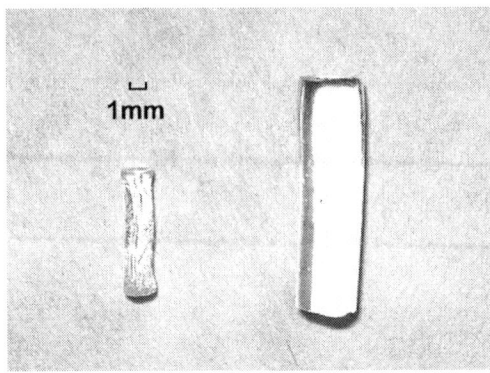

Figure 21. Hydrogel implants before (a) and after (b) 24 h in saline.

The resistance to the urine flow corresponds to the pressure of 20-25 cm of water column in the urinary bladder. Thus, the generated subvesical obstruction permits a spontaneous bladder emptying without residua. The implant is prepared from a strongly hydrophilic hydrogels, such as the copolymers of 2-hydroxyethyl methacrylate (HEMA) with sodium methacrylate [100], or acrylonitrile with acrylamide [101]. The clinical studies [102] showed the significant improvement of incontinence for 17 women (age ranging between 35 and 84 years), while about a half of them experienced a very significant improvement of incontinence.

Spherical hydrogel microparticles represent a type of modified hydrogel implants [103]. Their size ranges from 30 to 300 µm and they are applied in the dry state by injection into a designated location. After swelling, they create the demanded obstruction. However, the clinical effects tend to disappear after several months (in contrast to cylindrical implants mentioned above), either due to migration of the particles away from the injection site (caused by weak adherence with the surrounding soft tissues) or due to fibrosis (caused by excessive encapsulation of the particles by fibrous tissue). Little is known about the fate of injected microparticles, due to the fact that they are extremely difficult to trace in a noninvasive manner.

Chapter 7

Blood Vessel Embolization

Embolization of blood vessels has emerged as a highly effective technique in a wide variety of diseases. The method is based on keeping the embolic region away from blood circulation. But the efficiency of the method, and thus the degree of devascularization, was found to be limited by the development of a new vascular network around the pathologic area, which has the tendency to nourish the isolated region [104]. Such revascularization is associated with an inflammatory reaction of giant cells (macrophages), as it was observed by histological investigation [105]. The most promising results in clinical practice for the obliteration of blood vessels such as uterine arteries in the treatment of uterine fibroids have been obtained using calibrated, soft, compressible, and relatively hydrophilic microspheres of controlled shape and dimensions [106]. Since 1995 up to 2007, more than 50,000 cases of uterine fibroid embolizations, which are the most common tumors in the female genital tract, were performed worldwide [107]-[109]. Embolization is also used for treatment of inoperable tumors [110], arterio–venous malformations [111], craniofacial vascular malformation [112] and haemoptysis (excessive bleeding) [113],[114]. It is appropriate to modify embolization material by any roentgen-contrast compound in order to check the embol position roentgenologicaly.

Three ways of embolization are applied:

1. Use of a homogeneous, for example ethanolic solution of polymers (e.g. polyHEMA) insoluble in water, respectively in the blood [115]. When such solution is applied into the vessel, polymer precipitates

and obstructs the vessel. This system allows dissolving a roentgen-contrast substance in ethanolic solution of non-crosslinked polyHEMA which is insoluble in the blood and remains in embolus of polyHEMA. In similar manner calcium alginate can be used [116].

2. Use of the hydrogel microparticles dispersed in saline. The mechanical obstruction is formed upon the injection of microparticles into the vessel. The advantage of regular microspheres over irregularly shaped particles is that the uniform geometrical shape allows precise location of the embolic material. Further improvement of the microspheres can be achieved by biofunctionalization of the surface. By immobilization of a coagulation factor, such as thrombin, on the surface, the blood coagulation can be triggered. The advantage of this way for embolization therapies is that the spheres get anchored. Suitable hydrogel for microparticles proved to be the terpolymer of methyl methacrylate, methacrylic acid and 2-[4-iodobenzoyloxy]-ethyl methacrylate [117] crosslinked by the tetraethylene glycol dimethacrylate. Comonomer containing iodine serves as a roentgen-contrast agent, methacrylic acid allows the bounding of coagulation agent (thrombin) and methyl methacrylate improves the mechanical properties of hydrogel. Microparticles are prepared by dropwise addition of the mixture of monomers and initiator (benzoyl peroxide) to the aqueous solution of polyethylene glycol, polyvinyl alcohol and polyvinylpyridine under stirring at 85 C. After polymerization, the microparticles are washed with water and lyophilized. Thrombin is covalently bound by its amino group to the carboxylic group of a hydrogel. Due to the good adhesion of thrombin to the walls of a blood vessel of laboratory mice, compact embol is created and its position and shape in the vessel is easily radiologically observed. Crosslinked microparticles of poly(acrylamide) known under the commercial name of Trisacryl® can be used in a similar manner [118].

3. 3) Use of the metal wires, cylinders or spheres coated by a hydrogel [119]. This method is a modification of the previous method and its advantage is in the easier shape adjustability of the embolization implant

Chapter 8

Wound Dressings

Extensive burn injuries and other large skin defects of traumatic and/or metabolic nature are a serious medical, social and economic problem and the management of wound healing process still represents a challenge for current research in the medical area. The variety of wound types (burns, diabetic leg ulcers, pressure ulcers, surgical wounds, and etc.) has resulted in a wide range of possibilities of using different types of wound dressings to the successful management of the therapeutic process. Wound dressings have undergone a considerable evolution from crude application of plant herbs, over traditional dressings (cotton wool, natural and synthetic bandages and gauzes), that simply cover and conceal the wound, to modern materials ensuring the moist healing process, and more recently, to advanced tissue engineered scaffolds, which can replace the damaged skin.

Perhaps one of the most important advances to change the nature of wound dressing materials has been the confirmation of the importance of a moist environment around the wound to facilitate the healing outlined by Winter in 1962 [120]. Since then, the concept of the moist wound healing has been largely examined, which has led to the development of hundreds of modern wound dressings that support a moist wound environment, the results of which have been reviewed elsewhere [121]-[126]. Examples of commercial products available on the market today include hydrocolloids Granuflex™ (Conva Tec, UK), Comfeel™ (Coloplast, USA), or Tegasorb™ (3M Healthcare, UK), alginates Sorbsan™ and Tegagen™ (3M Healthcare, UK), and hydrogels NuGel™ (Johnson and Johnson), Purilon Gel™ (Coloplast, USA), or IntraSiteGel™ (Smith and Nephew).

Hydrogels play in the area of wound dressing an important role as they meet the most requirements for an "ideal" dressing due to their ability to contain significant amount of water (up to 90 wt.-%), that guarantees the maintaining of moist environment together with the possibility to absorb excess of exudates. Moreover, they are nonirritant and well tolerated with living tissue, permeable to metabolites, cause no trauma on removal due to their elastic character, and they cool the surface of the wound, which may lead to a marked reduction in pain and therefore high patient acceptability.

A detailed discussion about application of hydrogels in wound healing is, however, beyond the scope of this review. Therefore, in the following paragraphs we will focus on the methacrylate hydrogels applicable in wound healing being developed in our research group, i.e. hydrogel dressing materials containing radical scavengers and hydrogel supports suitable for keratinocyte cultivation for skin damage repair.

8.1. Hydrogels Containing Radical Scavengers

It is known that during the healing process a large number of different free radicals is formed in wound with the task of local disinfection. However, the free radicals besides protection from foreign micro-organisms also destroy the own cells and thus slow down the process of wound healing.

This negative aspect can be solved by incorporation of radical traps into the hydrogel structure. Such hydrogel dressing then shows the synergic effect of moist healing and deactivation of the free radicals. We have patented the preparation of slightly crosslinked hydrophilic polymer support based on methacrylic acid derivatives, in which biologically active compounds with radical catcher properties (derivatives of the vitamins A and E) were dispersed or dissolved in polymer matrices in amount up to 50 wt.-% [127]. It has been demonstrated that the wounds treated with this type of dressings healed without any secondary infection and successful epithelialisation was observed.

Another developed material belonging to group of products for the wound treatment is methacrylate gel containing in its structure covalently bonded sterically hindered amines, known for their ability to neutralize free radicals and thus facilitating the healing process [128]. This product is nowadays available on the market under the name HemaGel produced by Wake Pharma company.

8.2. Hydrogels as Cultivation Supports

The tissue-engineered skin represents a recent significant advancement in the wound healing. A number of products is nowadays commercially available; particularly Integra^{TM}, Alloderm^{TM}, or Apligraf^{TM} [121],[129] and many others are currently being developed.

The method of keratinocyte cultivation and subsequent transplantation was developed in the USA in 1979 [130]. It is predominantly used to re-epithelialisation of burns or wounds. Using this methodology, it is possible to prepare several thousand square centimetres of epidermal sheets in 3-4 weeks. Cultivation of autologous cells proceeds under the tissue cultures in the presence of 3T3 mice fibroblasts (feeder cells), in whose the proliferation was stopped by irradiation. The confluent growth of keratinocytes is detached from the bottom of cultured Petri dishes by the enzyme dispase and transferred on a vaseline gauze applied on the patient's wounds. The cell transfer is however technically demanding and it is often the cause of failure.

Therefore, the method allowing the direct cultivation of keratinocytes on a polymer support and its direct transfer to the wound was developed [131]. For the hydrogel support, the poly(2-hydroxyethyl methacrylate) (polyHEMA) was tested as a material with a long tradition in its biomedical use and a wide scale of applications. The process of keratinocyte cultivation is schematically shown in Figure 22. Firstly, polyHEMA supports were preincubated in bovine serum with a cocktail of proadhesive proteins (A) and then the lethally irradiated 3T3 mice fibroblasts were seeded to polyHEMA (B). After adhesion and covering the supports with feeder cell network, a suspension of keratinocytes was added (C). In seven to ten days keratinocytes formed a subconfluent to confluent growth and destroyed the feeder cells (D). The damaged feeder cells were removed by trypsinization and the support with grafted epidermal cells was directly applied in an upside-down position to the wound bed (E).

The results of clinical trials in a number of burned patients were highly encouraging [131],[132]: the cells, grafted upside down in a monolayer, migrated from the support and colonized the wound. This method, in comparison with classical procedure of the keratinocyte sheets on textile supports [130] is much easier; no enzymatic detachment of cells is necessary; the time necessary for the preparation of grafts is shorter, because the subconfluent growth of cells can be used; the hydrogel sheet on the surface of the wound bed protects the wound bed with transplanted cells and optimizes the microclimate.

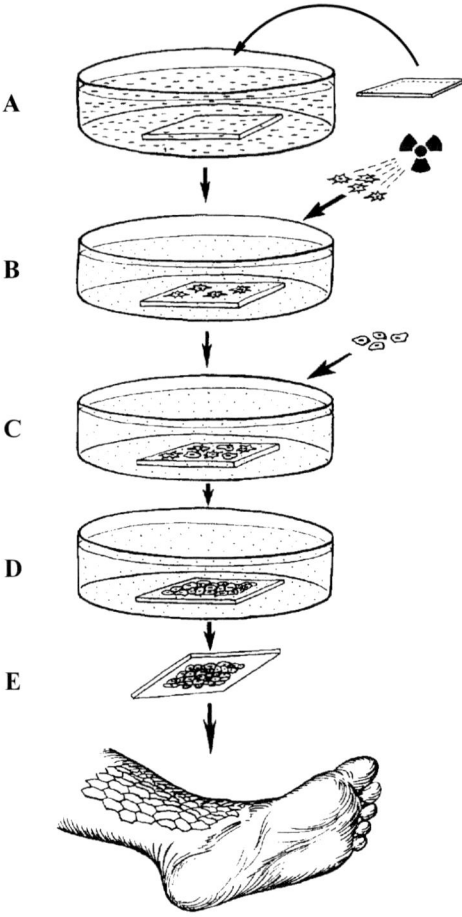

Figure 22. Scheme of keratinocyte cultivation on polyHEMA hydrogel supports and their application to the wound bed [131]. A – preincubation of polyHEMA in bovine serum, B – seeding of lethally irradiated 3T3 mice fibroblast, C – seeding of keratinocytes on the polymer support covered by network of feeder cells,
D - keratinocyte proliferation, E – direct application of the support to the wound bed.

Although the described upside-down technology brings desirable healing effect, the need of feeder cell used for initial attachment and growth of keratinocytes represents a certain complication form the point of view of clinical application. So, the exclusion of feeder cells would lead to the more effective process of keratinocyte cultivation, and moreover, the patient's immunologic burden would be reduced.

In the course of systematic study of biological properties of synthetic methacrylate hydrogels as potential supports for keratinocyte cultivation, we have observed that poly(2-ethoxyethyl methacrylate) (polyEOEMA) as a component of cultivation support stimulates the growth of human keratinocytes under in vitro conditions without feeder cells [133]. After 7 days, the surface of polyEOEMA was colonized with nearly confluent or confluent monolayer of cultured keratinocytes with extensive intercellular contacts. Moreover, keratinocytes cultured on these surfaces were able to migrate to the model wound bed in vitro, where they formed distinct colonies and had a normal differentiation potential confirmed by keratin immune-histochemistry.

This procedure is inexpensive and technically easy and therefore it seems to be suitable for keratinocyte culture in large scale and applicable for cell therapy of skin defects.

In recent years, advanced approach to eliminate the need of feeder cells for keratinocyte cultivation has received much attention, i.e. the preparation of cultivation supports modified by suitable bioactive pro-adhesive motifs, which would promote the adhesion, growth and proliferation of cells.

The preparation of bioactive polymer supports with covalently bound saccharides was patented [134]. Polymer is treated with a modification agent to activate the functional groups on the polymer surface (e.g. hydroxy, amino, carboxyl groups) and then it is allowed to react with appropriate derivative of saccharide (e.g. mannose, galactose, or lactose).

A desirable role of mannosides on the growth of human keratinocytes under in vitro conditions was recognized [135]. Mannosylated hydrogels were chemically synthesized and the successful cultivation of keratinocytes without feeder cells on such bioactive supports was demonstrated in *in vitro* experiments. These results offer a promising perspective; nevertheless the method is still too expensive for large-scale keratinocyte cultivation for the clinical practice.

It can be concluded that the modification of polymer supports by biologically active motifs is suitable for specific cell cultivation. So, the possibility of application of such polymer supports in the large-scale production will certainly be further investigated.

Chapter 9

Conductive Hydrogels for Biomedical Use

Electrical conductivity of the hydrogels can be achieved by introducing of any ionizable monomers as salts of acids (acrylic, methacrylic, itaconic, sulfopropylmethacrylate, vinylsulfonic, styrenesulfonic, aminoacids), or monomers with quaternary aminogroup (N,N,N-trimethylammoniumstyrene chloride, methacryloyloxyethylammonium chloride, N-methyl-3-vinyl-pyridinium chloride, [2-(methacryloyloxy)ethyl]dimethyl-(3-sulfopropyl)-ammonium chloride) into a polymer chain. Conductivity is a function of the amount of ionizable monomer in the polymer chain and of the acidobasic properties of the system [135]-[139]. The use of homopolymers containing ionizable monomers is usually unsuitable for their very high swelling degree in water (water content at equilibrium swollen hydrogel is often above 95 wt.-%) followed by poor mechanical properties. Another possibility is the use of non-ionizable hydrogels, in which some conductive particles (salts, metals) are dispersed either touching each other or the contact between particles must be mediated through a conductive liquid electrolyte swollen into hydrogel [140].

In biomedical field, the conductive hydrogels are used mainly as biosensors [138]-[139],[141]-[143], ECG, EEG, defibrillation or high-frequency electrodes for electrosurgery, materials for tissue engineering [144] or for the enzymes detection [145]. In this chapter we will focus only on the application in electrodes.

Figure 23. Adhesive conductive hydrogel on ECG electrode.

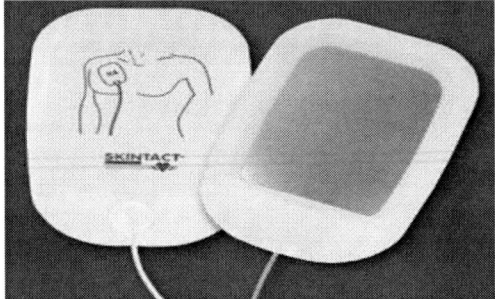

Figure 24. SKINTACT Easibeat Multifunction Defibrillation Electrodes designed for the use of defibrillation, non-invasive pacing/synchronized cardio-version and monitoring.

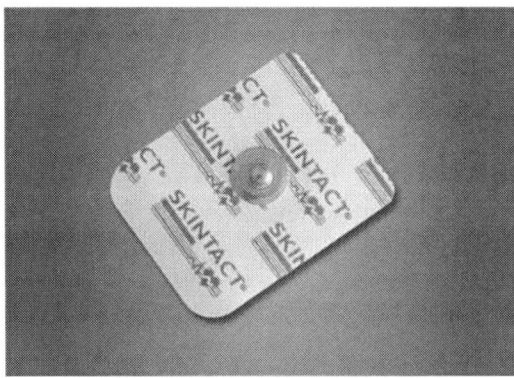

Figure 25. SKINTACT ECG electrode.

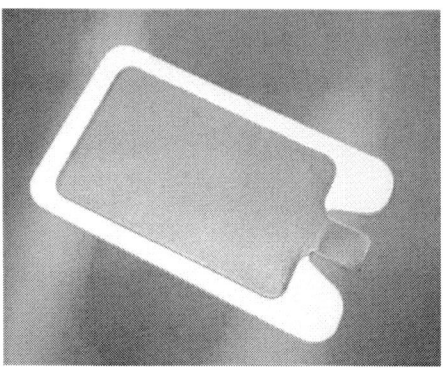

Figure 26. SKINTACT Electrode for electrosurgery.

Electrodes based on hydrogels are used in medicine predominantly for electrosurgery [146],[147], ECG and EEG [148] measurements and defibrillation [149]. In these applications, the use of conductive hydrogels with a good adhesion to the skin is usually required (cf. Figure 23) [150].

Such a hydrogel can be prepared by crosslinking polymerization, initiated by UV radiation or oxidative-reductive system, of moderately hydrophilic monomers such as the 2- hydroxyethyl methacrylate (HEMA), in combination with the ionogenic sodium methacrylate and in the presence of water and non-volatile hydrophilic solvents such as polyethylene glycols or glycerol. Conductivity is affected by the presence of sodium methacrylate dissociated in water. The adhesion power is given by the presence of hydrophilic solvent in sub-equilibrium concentration compare to free equilibrium swollen hydrogel; this guarantees that no components of hydrogel remain on the skin after removal of the electrode. Electrode is built from a metal foil connected to a diagnostic device (ECG, EEG), or in other instances to the source of electrical power, adhesive conductive hydrogel layer and protective foil that prevents hydrogel drying during the storage. The foil is removed immediately before use and the hydrogel layer is placed directly on the skin. Typical examples of such electrodes are products of Leonhard Lang company [150], named SKINTACT ® (Figure 24-Figure 26).

Chapter 10

Hydrogels in Tissue Engineering

The area of tissue engineering (with cell therapy, drug delivery systems, and genetic engineering) is one of the most advancing biomedicinal disciplines especially because it offers treatment options where traditional medicine fails. The hydrogel materials play in this field a key role: as they enable the preparation of variety of scaffolds essential in tissue engineering. As this area is very wide, in this paragraph we will focus only on the applications of hydrogels in the central nervous system (CNS) and in particular in the spinal cord.

Brain or spinal cord injury causes severe tissue damage resulting in a permanent neurological deficit. Spontaneous regeneration is severely limited. The glial scar, mesenchymal scar and posttraumatic pseudocysts form an obstacle for regeneration. Tissue engineering continues to assume greater importance in neuroscience, with the ultimate goal of restoring the morphology and function of damaged nervous tissue. Tissue engineering techniques help to create a permissive environment to promote the regeneration of neurons and axons, oligodendrocytes, and blood vessels.

Various biomaterials, including hydrogels, have been implanted inside a CNS in order to bridge the lesion [151],[152]. Hydrogels have many advantages over some alternative scaffold materials, such as high oxygen and nutrient permeability and low interfacial tensions. The latter attribute minimizes barriers to cell migration into the scaffold from the surrounding soft tissue. The hydrogel scaffolds used in neural tissue engineering are commonly macroporous (typical morphology is shown on Figure 27) or become macroporous upon degradation, to allow room for neurite outgrowth. From this point of view, the pores must be communicating or prevailingly communicating as was described before [153]. Therein, other characteristics regarding morphology as total porosity, pore size, etc. are also referred. Further, the influence of both, positive and negative charge incorporated into polymer network on resulting structure of macroporous hydrogels was studied and reported in [154]. Preparation of macroporous hydrogels based on various methacrylate monomers (e.g. HEMA, HPMA, EOEMA) using pH sensitive and hydrolytically degradable N,O-dimethacryloylhydroxylamine as a crosslinker was studied and the possibility of adjusting the degradation time between 2 and 42 days was showed [155].

The mechanical properties of hydrogels are similar to those of soft tissue so that the load on cells in the area will be distributed normally and cells within the gel receive appropriate structural growth cues. The elasticity of hydrogels is also tunable by controlling the macromolecular structure and crosslink density. Another advantage of many hydrogels is that they easily conform to any defect shape, whilst they can be functionalized to include neurotrophins for the control of neuronal cell adhesion, proliferation and axonal extension. Cell replacement therapies in the CNS may also greatly benefit using hydrogels that provide artificial 3D stem cell niches for controlled proliferation and differentiation - an area of considerable activity in regenerative medicine.

Figure 27. SEM micrograph of the typical morphology of macroporous hydrogel based on EOEMA/HPMA.

Diffusion parameters within implanted hydrogels attain values similar to those of developing neural tissue [156]. The physical and chemical properties of hydrogels can be modified to improve cell adhesion and tissue regeneration. Further, the synthetic hydrogels can be produced in large quantities, and combined with allogeneic or autologous transplants. Previous studies of ours and others have shown that hydrogels based on HEMA are promising biomaterials for CNS regeneration [151],[153]-[158]. Increasing knowledge of the pathophysiology of CNS, cell adhesion properties, molecular biology, and biomaterial science could lead to the development of implants that would successfully bridge a spinal cord lesion and lead to a complete recovery of locomotor, sensory, and autonomic functions. It is well known that the physical and chemical properties of a surface can influence cellular adhesion [159]. Previous studies have shown that neurons preferentially adhere to and form neural networks on positively charged surfaces such as polylysine-coated glass slides [160]. In our previous study, we found that HEMA-based hydrogels with positively-charged functional groups promote the adhesion of mesenchymal stem cells [157] that promote protection against tissue damage in experimental SCI and can increase the expression of growth and trophic factors in the ischemic rat brain.

The Figure 28 shows the tissue infiltration in the four HEMA-based hydrogels with different charge [161]. In the negatively charged hydrogels (HEMA-, copolymer HEMA with sodium methacrylate), there was only a small amount of connective tissue in the hydrogels visible on hematoxylin-eosin stained sections (A, B). The amount of connective tissue elements increased in hydrogels with both positive and negative charges (HEMA±, i.e. the terpolymer of HEMA with methacryloyloxyethyl-trimethylammonium chloride and with sodium methacrylate) (E, F) and was highest in the positively charged hydrogels (HEMA+, copolymer of HEMA with methacryloyloxyethyltrimethylammonium chloride) (I, J). In the polyelectrolyte complex (PEC, complex of poly(sodium methacrylate) with poly(methacryloyloxyethyltrimethylammonium chloride)) implants, minimal connective tissue elements infiltrated the implant (M, N). The pseudocystic cavity dominated the lesion after hemisection without hydrogel implantation (Q), with a sharp border between the pseudocystic cavity and the residual tissue (R). Neurofilaments (arrows) were found in hydrogels with functional groups (C, G, K) in contrast to PEC implants (O). We found axons infiltrating the peripheral parts of the implants in all hydrogels with functional groups (C), while hydrogels with positive charges (G) had many axons also infiltrating the central parts of the implants. Few astrocytic processes were found especially in the HEMA- (D) and PEC implants (P) as compared with minimal astrocytic processes in the HEMA± (H) and HEMA+ (L). The astrocytes formed a glial scar (white arrow) around the hemisection cavity (S). Hydrogels are marked with an asterisk.

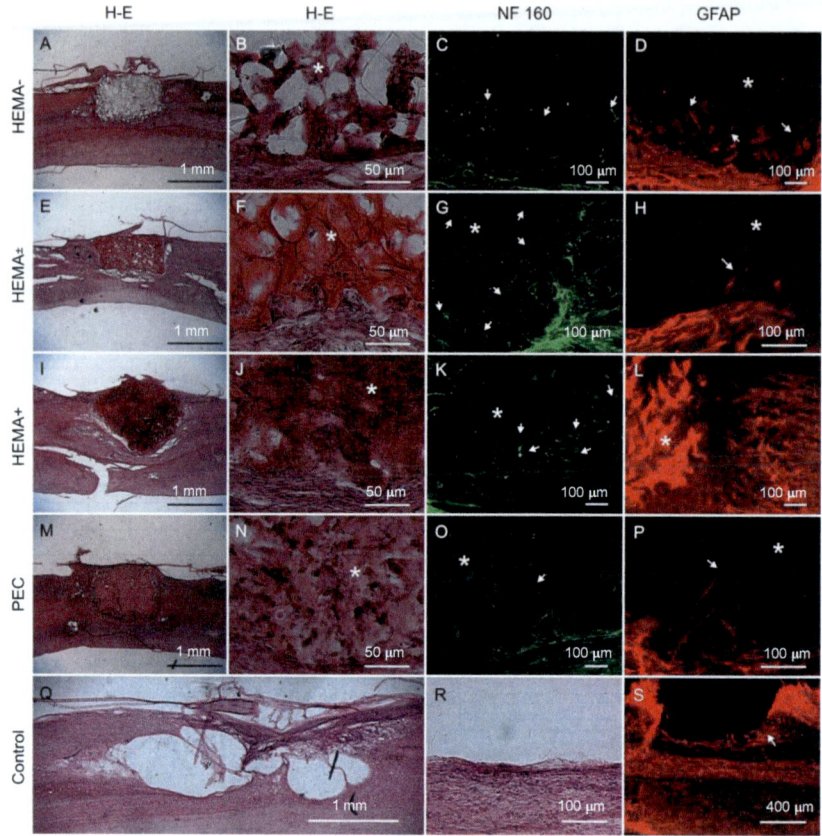

Figure 28. Tissue infiltration in the four HEMA-based hydrogels with different charges [161].

References

[1] Almdal, K; Dyre, J; Hvidt, S; Kramer, O. What is a gel? *Makromol. Chem. Macromol. Symp.*, 1993 76, 49-51.

[2] Kunzler, JF. *Enclyclopedia of polymer science*. New York: J. Wiley; 2002.

[3] te Nijenhuis, K. Thermoreversible networks. Viscoelastic properties and structure of gels. *Adv. Polym. Sci,.* 1997 130, 1-193.

[4] Dusek, K, editor. Responsive gels. Volume phase transition. *Adv. Polym. Sci.*, 1993, Vol. 109; 1993, Vol. 110, Springer, Berlin.

[5] Dusek, K; Duskova-Smrckova, M. Network structure formation during crosslinking of organic coating systems. *Prog. Polym. Sci.*, 2000 25, 1215-1260.

[6] Dusek, K; Duskova-Smrckova, M. Structure-property correlation and characterization techniques. In: Matyjaszewski, K; Gnanou, P; Leibler, L, editors. *Polymer networks in macromolecular engineering. Precise synthesis, materials properties, applications.* Weinheim: Wiley-VCH; 2007; 1687-1730.

[7] Winter, HH; Mourns, M. Rheology of polymers near liquid-solid transitions. *Adv. Polym. Sci.*, 1997 134, 165-234.

[8] Allain, C; Drifford, M; Gauthier-Manuel, B. Diffusion of calibrated particles during the formation of a gel. *Polymer Commun.*, 1986 27, 177-180.

[9] Norisuye, T; Shibayama, M; Nomura, S. Time-resolved light scattering study on the gelation process of poly(N-isopropylacrylamide). *Polymer*, 1998 39, 2769-2775.

[10] Larsen, TH; Furst, EM. Microrheology of the liquid-solid transition during gelation. *Phys. Rev. Lett.*, 2008 100, 146001-1 – 146001-4.

[11] Polymer networks - principles of their formation, structure and properties, Stepto, RFT, editor. London: Blackie Academics and Professional; 1998.
[12] Dusek, K; Duskova-Smrckova, M; Yang, J; Kopecek, J. Coiled-coil hydrogels: Effect of grafted copolymer composition and cyclization on gelation. *Macromolecules*, 2009 42, 2265-2274.
[13] Lodge, TP. A virtual issue of macromolecules "Click chemistry in macromolecular science". *Macromolecules*, 2009 42, 3827-3829.
[14] Bomgarden, R. Studying protein interactions in living cells; method for protein crosslinking utilizes photoreactive amino acids. *GEN*, 2008 28(7), 24-25.
[15] Pradny, M; Michalek, J; Sirc, J. Porous hydrogels. In: Acosta, JL; Camacho, AF, editors. *Porous media: heat and mass transfer, transport and mechanics*. New York: Nova Publishers; 2009; 57-74.
[16] Dusek, K; Prins, W. Structure and elasticity of non-crystalline polymer networks. *Adv. Polym. Sci.*, 1969 6, 1-102.
[17] Duskova-Smrckova, M; Dusek, K; Vlasak, P. Solvent activity changes and phase separation during crosslinking of coating films. *Macromol. Symp.*, 2003 198, 259-270.
[18] Dusek, K; Sedlacek, B. Phase separation in poly(2-hydroxyethyl methacrylate) gels in the presence of water. *Eur. Polym. J.*, 1971 7, 1275-1285.
[19] Flory, PJ. *Principles of polymer chemistry.* Ithaka: Cornell University Press; 1953.
[20] Dusek, K; Prins, W. Structure and elasticity of non-Crystalline polymer networks. *Adv. Polym. Sci.*, 1969 6, 1-102.
[21] Erman, B; Mark, JE. *Structure and properties of rubberlike networks.* Oxford: Oxford University Press; 1997.
[22] Koningsveld, R; Stockmayer, WH; Nies, E. *Polymer phase diagrams. A textbook.* Oxford: Oxford University Press; 2001.
[23] ten Brinke, G; Karasz, FE. Lower critical solution temperature behavior in polymer blends: compressibility and directional-specific interactions. *Macromolecules*, 1984, 17, 815-820
[24] Hino, T; Labert, SM; Soane, DS; Prausnitz, JM. Miscibilities in binary copolymer systems *Polymer*, 1993, 34, 4756-4761.
[25] Solc, K; Dusek, K; Koningsveld, R; Berghmans, H. "Zero" and "off-zero" critical concentrations in solutions of polydisperse polymers with very high molar masses. *Collect. Czech. Chem. Commun.*, 1995 60, 1661-1668.

References

[26] Moerkerke, R; Koningsveld, R; Berghmans, H; Dusek, K; Solc, K. Phase transitions in swollen networks. *Macromolecules*, 1995 28, 1103-1107.

[27] Dusek, K. Quasichecal approach to crosslinked polymer solutions and swelling equations for polycondensation networks. *J. Polym. Sci, Polym. Phys. Ed.*, 1974 12, 1089-1107.

[28] Dusek, K; Sedlacek, B. Effect of diluents on the microsyneresis in poly(2-hydroxyethyl methacrylate) gels induced by temperature changes. *Collect. Czech. Chem. Commun.*, 1971 35, 1567-1577.

[29] Janacek, J; Hasa J. Structure and propertiers of hydrophilic polymers and gels VL. Equilibrium deformation behaviour of polyethylene glycol methacrylate and polyethyleneglycol methacrylate networks prepared in the presence of a diluent and swollen with water. *Collect. Czech. Chem. Commun.*, 1965 31, 2186-2201.

[30] Rodriguez, O; Fornasiero, F; Arce, A; Radke, CJ; Prausnitz, JM. Solubilities and diffusivities of water vapor in poly(methyl methacrylate), poly(2-hydroxyethyl methacrylate, poly(N-vinyl-pyrrolidone) and polyacrylonitrile. *Polymer*, 2003 44, 6323-6333.

[31] Weinmuller, C; Langel, C; Radke, CJ; Prausnitz, JM. Sorption kinetics and equilibrium uptake for water vapor in soft contact lens hydrogels. *J. Biomed. Mater. Res.*, 2006 77A, 230-241.

[32] Fornasiero, F; Ung, M; Radke, CJ; Prausnitz, JM. Glass transition temperatures for soft-contact-lens materials. Dependence of water content. *Polymer*, 2005 46, 4845-4852.

[33] Hasa, J; Ilavsky, M; Dusek, K. Deformational swelling and potentiometric behaviour of ionized polymethacrylic acid gels. *J. Polym. Sci., Polym. Phys. Ed.*, 1975 13, 253-262.

[34] Victorov, AI; Radke, CJ; Prausnitz, JM. Molecular thermodynamics for swelling of a bicontinuous gel. *Mol. Phys.*, 2002 100, 2277-2297.

[35] Dusek, K; Duskova-Smrckova, M; Ilavsky, M; Stewart, R; Kopecek, J. Swelling pressure induced phase-volume transition in hybrid biopolymer gels caused by unfolding of folded domains: A model. *Biomacromolecules*, 2003 4, 1818-1826.

[36] Dusek, K; Patterson, D. A transition in swollen polymer networks induced by intramolecular condensation. *J. Polym. Sci., Part. A-2*, 1968 6, 1209-1216.

[37] Tanaka, T. Collapse of gels and critical endpoint. *Phys. Rev. Lett.*, 1978 40, 820-823.

[38] Schild; HG. Poly (N-isopropylacrylamide) - experiment, theory and application. *Prog. Polym. Sci.*, 1992 17, 163-249
[39] Okano, T. Biorelated polymers and gels: Controlled release and applications in biomedical engineering. Academic Press; 1998.
[40] Tompa, H. *Polymer Solutions.* London: Butterworth; 1956.
[41] Dusek, K. Phase separation in ternary systems induced by crosslinking. *Chem. Zvesti*, 1971 25, 184-188.
[42] Dusek, K. Parameters of the swelling equation and network structure. *Faraday Disc. Chem. Soc.*, 1974 57, 101-109.
[43] Dusek, K. Phase separation during the formation of three-dimensional polymers. *J. Polym. Sci. Part. C*, 1967 16, 1289-1299.
[44] Dusek, K; Duskova-Smrckova, M. Vapor pressure over stressed coatings. *Polym. Bull.*, 2000 45, 83-88.
[45] Mark, JE; Erman B. *Rubberlike elasticity. A molecular primer.* Wiley, New York; 1988.
[46] Cluff, EF; Gladding, EK; Pariser, R. A New method for measuring the degree of crosslinking in elastomers. *J. Polym. Sci.* 1960 45, 341-345.
[47] Chambon, F; Winter, HH. Stopping of crosslinking reaction in a PDMS polymer at the gel point. *Polym. Bull.* 1985 13, 499-504.
[48] Ferry, JD. *Viscoelastic properties of polymers.* Wiley. New York, 1st ed 1961, 2nd ed 1970, 3rd ed 1980.
[49] Macosko, CW. Rheology. Principles, Measurements, and Applications. Wiley-VCh, 1994.
[50] Pascault, JP; Sauterau, H; Verdu, J; Williams, RJJ. *Thermosetting polymers.* Marcel Dekker, Basel, Switzerland, 2002.
[51] Waigh, TA. Microrheology of complex fluids. *Rep. Prog. Phys.* 2005 68, 685-742.
[52] http://www.seas.harvard.edu/weitzlab/research/micrheo.html
[53] Levine, AJ; Bubensky, TC. One- and two- particle microrheology. *Phys. Rev. Lett.* 2000 85, 1774-1777.
[54] Gittes, F; Schnurr, B; Olmsted, PD; MacKintosh, FC; Schmidt, CF. Microscopic viscoelasticity: Shear moduli of soft materials determined from thermal fluctuations. *Phys. Rev. Lett.* 1997 79, 3286-3289.
[55] Discher, DE; Janmey, P; and Wang, YL. Tissue cells feel and respond to the stiffness of their substrate. *Science* 2005 310, 1139-1143.
[56] Helmlinger, G; Netti, PA; Lichtenbeld, HC; Melder, RJ; Jain, RK. Solid stress inhibits the growth of multicellular tumor spheroids. *Nat. Biotechnol.* 1997 15, 778-783.

[57] Yu, TT; Shoichet, MS. Guided cell adhesion and outgrowth in peptide-modified channels for neural tissue engineering, *Biomataterials* 2005 26, 1507-1514.

[58] Anseth, KS; Bowman, CN; Brannon-Peppas, L. Mechanical properties of hydrogels and their determination. *Biomaterials* 1996 17, 1647-1657.

[59] Heitz, RF; Enoch, JM. Leonardo da Vinci: An assessment on his disclosures on image formation in the eye. In: Fiorentini, A; Guyton, DL; Siegel, IM, Editors. *Advances in diagnostic visual optics*. Springer-Verlag; 1987; 19–26.

[60] Enoch, JM. Descartes´contact lens. *Am. J. Optom. Arch. Am. Acad. Optom.*, 1956 33, 77-85.

[61] Young, T. On the mechanism of the eye. *Phil. Trans. R. Soc. Lond. Biol. Sci.*, 1801 91, 23-88.

[62] www.eyetopics.com. The history of contact lenses. Accessed October 18, 2006.

[63] Efron, N; Pearson, RM. Centenary celebration of Fick's Eine Contactbrille. *Arch. Ophthalmol.*, 1988 106, 1370-1377.

[64] Pearson, RM. Kalt, keratoconus and contact lens. *Optom. Vis. Sci.*, 1989 66, 643-646.

[65] Pearson, RM; Efron, N. Hundredth anniversary of August Muller's inaugular dissertaion on contact lenses. *Surv. Ophthalmol.*, 1989 34, 133-141.

[66] Feinbloom, W. A plastic contact lens. *Trans. Am. Acad. Optom.*, 1936 10, 37-44.

[67] Tuohy, KM. Contact lens. US Patent 2510438, Jun 1950.

[68] Tighe, B. Rigid lens materials. In: Efron, N, editor. *Contact Lens Practice*. Butterworth-Heinemann; 2002; 153-162.

[69] Gaylord, NG. Oxygen-permeable lens composition, methods and article of manufacture. US Patent 3808178, Apr 1974.

[70] Wichterle, O; Lim, D. Hydrophilic gels for biological use. *Nature*, 1960 185, 117-118.

[71] Wichterle, O; Lim, D. Process for producing shaped articles from three-dimensional hydrophilic high polymers. US Patent 2976576, March 1961.

[72] Tighe B. Soft lens materials. In: Efron, N, editor. *Contact Lens Practice*. Butterworth-Heinemann; 2002; 71-84.

[73] Nilsson, S. Ten years of disposable lenses – a review of benefits and risks. *Contact Lens Ant. Eye*, 1997 20, 119-128.

[74] Tighe, B. Silicone hydrogel materials – how do they work? In: Sweeney, D, editor. *Silicone hydrogels: The rebirth of continuous wear contact lenses*. Oxford: Butterworth-Heinemann; 2000, 1-21.
[75] O´Brien, TP. Advances in intraocular lens materials and designs: Maximizing biocompatibility and optical performance. *Ophthalmologica*, 2003 217, 7-18.
[76] Olson, RL; Werner, L; Mamalis, N; Cionni, R. New Intraocular lens technology. *Am. J. Ophthalmol.*, 2005 140, 709-716.
[77] Kanski, JJ. Clinical Ophthalmology. A systematic approach. Butterworth-Heinemann; 1989.
[78] Christ, FR; Fencil, DA; Van Gent, S; Knight, PM. Evaluation of the chemical, optical, and mechanical properties of elastomeric intraocular lens materials and their clinical significance. *J. Cataract. Refract. Surg.*, 1989 15, 176–184.
[79] Christ, FR; Buchen, SY; Deacon, J; Cunanan, M; Giamporcaro, JE; Knight, PM; Weinschenk, JI; Yang, S. Biomaterials used for intraocular lenses. In: Wise, DL, editor. *Encyclopedic handbook of biomaterials and bioengineering. Part B: Applications.* Marcel Dekker; 1995; Vol. 2, 1261–1313.
[80] Michalek, J; Vacik, J. Process for producing implantable intraocular planar/convex, biconvex, planar/concave or convex/concave lens, open mold for making such process and lens produced in such a manner. CZ Patent 297180.
[81] Werner, L. Biocompatibility of intraocular lens materials. *Curr. Opin. Ophthalmol.*, 2008 19, 41-49.
[82] Kang, S; Kim, MJ; Park, SH; Joo, CK. Comparison of clinical results between heparin surface modified hydrophilic acrylic and hydrophobic acrylic intraocular lens. *Eur. J. Ophthalmol.*, 2008 18, 377-383.
[83] Kleinmann, G; Apple, DJ; Chew, J; Stevens, S; Hunter, B; Larson, S; Mamalis, N; Olson, RJ. Hydrophilic acrylic intraocular lens as a drug-delivery system: Pilot study. *J. Cataract. Refract. Surg.*, 2006 32, 652–654.
[84] Dewey, S. Posterior capsule opacification. *Curr. Opin. Ophthalmol.*, 2006 17, 45–53.
[85] Abela-Formanek, C; Amon, M; Schild, G; Schauersberger, J; Kolodjaschna, J; Barisani-Asenbaum, T; Kruger, A. Inflammation after implantation of hydrophilic acrylic, hydrophobic acrylic, or silicone intraocular lenses in eyes with cataract and uveitis: Comparison to a control group. *J. Cataract. Refract. Surg.*, 2002 28, 1153–1159.

[86] Stoy, VA; Pasta, J; Novak, J; Vesely, P. Resistance of WIOL-CF hydrogel IOL to calcification. Best Paper of Session (BPOS) Winners, Symposium on Cataract, IOL and Refractive Surgery, March 18-22, San Francisco, USA, 2006.
[87] Werner, L. Causes of intraocular lens opacification or discoloration. *J. Cataract. Refract. Surg.*, 2007 33, 713-726.
[88] Taboada-Esteve, JF; Hurtado-Sarrio, M; Duch-Samper, AM; Cisneros-Lanuza, A; Menezo-Rozalen, JL. Hydrophilic acrylic intraocular lens clouding: A clinicopathological review. *Eur. J. Ophthalmol.*, 2007 17, 588-594.
[89] Neuhann, IM; Stodulka, P; Werner, L; Mamalis, N; Pandey, SK; Kleinmann, G; Apple, DJ. Two opacification patterns of the same hydrophilic acrylic polymer: Case reports and clinicopathological correlation. *J. Cataract. Refract. Surg.*, 2006 32, 879–886.
[90] Smetana, K; Vacik, J; Souckova, D; Pitrova S. The influence of chemical functional groups on implant biocompatibility. *Clin. Mater.*, 1993 13, 47-49.
[91] Menapace, R; Findl, O; Kriechbaum, K; Leydolt-Koeppl, C. Accommodating intraocular lenses: a critical review of present and future concepts. *Graefe's Arch. Clin. Exp. Ophthalmol.*, 2007 245, 473-489.
[92] Schwartz, DM. Light-adjustable lens. *Trans. Am. Ophthalmol. Soc.*, 2003 101, 411-430.
[93] Levy, R; Muller, N. Urinary incontinence: economic burden and new choices in pharmaceutical treatment. *Adv. Ther.*, 2006 23, 556-573.
[94] Lotenfoe, R; O'Kelly, JK; Helal, M; Lockhart JL. Periure thral polytetrafluoroethylene paste injection in incontinent female subjects: surgical indications and improved surgical technique. *J. Urol.*, 1993 149, 279-282.
[95] Aboseif, SR; O'Connell, HE; Usui, A; Mc Quire, EJ. Collagen injection for intrinsic sphincter deficiency in men. *J. Urol.*, 1996 155, 10–13.
[96] Harris, DR; Iacovou, JW; Lemberger, RJ. Peri-urethral silicone microimplants for the treatment of genuine stress incontinence. *Br. J. Urol.*, 1996 78, 722–728.
[97] Madjar, S; Jacoby, K; Giberti, C; Wald, M; Halachmi, S; Issaq, E; Moskovitz, B; Beyar, M; Nativ, O. Bone anchored sling for the treatment of post-prostatectomy incontinence. *J. Urol.*, 2001 165, 72–76.
[98] Pfizer, AMS. US Patents: 3863622; 4222377, May 1991.

[99] Sankar, S; Rajalakshmi, T. Application of poly ethylene glycol hydrogel to overcome latex urinary catheter related problems. *Biofactors*, 2007 30, 217-225.
[100] Sefc, L; Pradny, M; Vacik, J; Michalek, J; Povysil, C; Vitkova I; Halaska, M; Simon, V. Development of hydrogel implants for urinary incontinence treatment. *Biomaterials*, 2002 23, 3711-3715.
[101] Ramseyer, P; Meagher-Villemure, K; Burki, M; Frey, P. (Poly)acrylonitrile-based hydrogel as a therapeutic bulking agent in urology. *Biomaterials*, 2007 28, 1185–1190.
[102] Lose, G; Mouritsen, L; Nielsen, JB. A new bulking agent (polyacrylamide hydrogel) for treating stress urinary incontinence in women. *BJU Int.*, 2006 98, 100-104.
[103] Saralidze, K; Knetsch, MLW; van Hooy-Corstjens, CSJ; Koole, LH. Radio-opaque and surface-functionalized polymer microparticles: Potentially safer biomaterials for different injection therapies. *Biomacromolecules*, 2006 7, 2991-2996.
[104] Gobin, YP; Vinuela, F; Vinters, HV; Ji, C; Chow, K. Embolization with radiopaque microbeads of polyacrylonitrile hydrogel: Evaluation in swine. *Radiology*, 2000 214, 113-119.
[105] Laurent, A; Wassef, M; Drouet, L; Pignaud, G; Merland, JJ. Etude histologique de plusieurs matériaux d'embolisation et d'un nouveau type de matériel sphérique et adhésif. *Innov. Tech. Biol. Med.*, 1989 10, 358-366.
[106] Pelage, JP; Le Dref, O; Beregi, JP; Nonent, M; Robert, Y; Cosson, M; Jacob, D; Truc, JB; Laurent, A; Rymer, R. Limited uterine artery embolization with tris-acryl gelatin microspheres for uterine fibroids. *J. Vasc. Interv. Radiol.*, 2003 14, 15–20.
[107] Baakdah, H; Tulandi, T. Uterine fibroid embolization. *Clin. Obstet. Gynecol.*, 2005 48, 361-368.
[108] Siskin, GP; Englander, M; Stainken, BF; Ahn, J; Dowling, K; Dolen, EG. Embolic agents used for uterine fibroid embolization. *Am. J. Roentgenol.*, 2000 175, 767-773.
[109] Lupattelli, T; Basile, A; Garaci, FG; Simonetti, G. Percutaneous extraluminal recanalization: usefulness of false channel balloon dilation and heparin administration before true lumen reentry. *Eur. J. Radiol.*, 2005 54, 136–147.
[110] Bendszus, M; Martin-Schrader, I; Warmuth-Metz, M; Hofmann E; Solymosi, L. MR-imaging—and MR spectroscopy—revealed changes in

meningiomas for which embolization was performed without subsequent surgery. *Am. J. Neuroradiol.*, 2000 21, 666-669.

[111] Fleetwood, I; Steinberg, GK. Arteriovenous malformations. *Lancet*, 2002 359, 863-873.

[112] Bodhey, NK; Gupta, AK; Purkayastha, S; Kesavadas, C; Krishnamoorthy, T; Kapilamoorthy, T; Thomas, B. Embolization of craniofacial vascular malformations. *Riv. Neuroradiol.*, 2005 18, 349-356.

[113] Yoon, W. Embolic agents used for bronchial artery embolization in massive haemoptysis. *Expert Opin. Pharmacother.*, 2004 5, 361-367.

[114] Vinaya, KN; White Jr., RI; Sloan, JM. Reassessing bronchial artery embolotherapy with newer spherical embolic materials. *J. Vasc. Interv. Radiol.*, 2004 15, 304-305.

[115] Pradny, M; Michalek, J, unpublished data.

[116] Becker, TA; Preul, MC; Bichard, WD; Kipke, DR; McDougall, CG. Calcium alginate gel as a biocompatible material for endovasculat arteriovenous malformation embolization: Six-month results in an animal model. *Neurosurg.*, 2005 56, 793-800.

[117] Saralidze, K; Van Hooy-Corstjens, CSJ; Koole, LH; Knetsch, MLW. New acrylic microspheres for arterial embolization: Combining radiopacity for precise localization with immobilized thrombin to trigger local blood coagulation. *Biomaterials*, 2007 28, 2457–2464.

[118] Madani, F; Bessodes, M; Lakrouf, A; Vauthier, C; Daniel Scherman, D; Chaumei, JC. PEGylation of microspheres for therapeutic embolization: Preparation, characterization and biological performance evaluation. *Biomaterials*, 2007 28, 1198-1208.

[119] Hiraki, T; Pavcnik, D; Uchida, BT; Timmermans, HA; Yin, Q; Wu, RH; Niyyati, M; Keller, FS; Rösch, J. Prophylactic residual aneurysmal sac embolization with expandable hydrogel embolic devices for endoleak prevention: Preliminary study in dogs. *Cardiovasc. Intervent. Radiol.*, 2005 28, 459-466.

[120] Winter, GD. Formation of the scab and the rate of epithelialization of superficial wounds in the skin of the young domestic pig. *Nature*, 1962 193, 293-294.

[121] Boateng, JS; Matthewa, KH; Stevens, HNE; Eccleston, GM. Wound healing dressings and drug delivery systems: A review. *J. Pharm. Sci.*, 2008 97, 2892-2923.

[122] Fonder, MA; Lazarus, GS; Cowan, DA; Aronson-Cook, B; Kohli, AR; Mamelak, AJ. Treating the chronic wound: A practical approach to the

care of nonhealing wounds and wound care dressings. *J. Am. Acad. Dermatol.*, 2008 58, 185-206.
[123] Ovington, LG. Advances in wound dressings. *Clin. Dermatol.*, 2007 25, 33-38.
[124] Heyneman, A; Beele, H; Vanderwee, K; Defloor, T. A systematic review of the use of hydrocolloids in the treatment of pressure ulcers. *J. Clin. Nurs.*, 2008 17, 1164-1173.
[125] Qin, Y. Review: Alginate fibres: An overview of the production processes and application in wound management. *Polym. Int.*, 2008 57, 171-180.
[126] Jones, A; Vaughan, D. Hydrogels dressings in the management of a variety of wound types: A review. *J. Orthop. Nurs.*, 2005 Suppl. 1, S1-S11.
[127] Michalek, J; Novak, P; Straskraba, I; Vacik, J; Wirthova, E. A wound-cover material containing radical scavengers. WO Patent 2004096367, Nov 2004.
[128] Labsky, J; Vacik, J; Hosek, P. Preparation for prevention and healing of inflammation affections. US patent: 6610284, Aug 2003.
[129] Ramos-E-Silva, M; de Castro, MCR. New dressings, including tissue-engineered living skin. *Clin. Dermatol.*, 2002 20, 715-723.
[130] Green, H; Kehinde, O; Thomas, J. Growth of cultured human epidermal cells into multiple epithelia suitable for grafting. *Proc. Natl. Acad. Sci. USA*, 1979 76, 5665-5668.
[131] Dvorankova, B; Smetana, K, Jr; Konigova, R; Singerova, H; Vacik, J; Jelinkova, M; Kapounkova, Z; Zahradnik, M. Cultivation and grafting of human keratinocytes on a poly(hydroxyethyl methacrylate) support to the wound bed: A clinical study. *Biomaterials*, 1998 19, 141-146.
[132] Dvorankova, B; Holikova, Z; Vacik, J; Konigova, R; Kapounkova, Z; Michalek, J; Pradny, M; Smetana, K, Jr. Reconstruction of epidermis by grafting of keratinocytes cultured on polymer support – clinical study. *Int. J. Dermatol.*, 2003 42, 219-223.
[133] Vacik, J; Dvorankova, B; Michalek, J; Pradny, M; Krumbholcova, E; Fenclova, T; Smetana, K, Jr. Cultivation of human keratinocytes without feeder cells on polymer carriers containing ethoxyethyl methacrylate: In vitro study. *J. Mater. Sci.: Mater. Med.*, 2008 19, 883-888.
[134] Labsky, J; Vacik, J; Smetana, K; Dvorankova, B. Polymer carrier for cultivation keratinocytes with active saccharides. CZ Patent 9901946, Jan 2001.

[135] Labsky, J; Dvorankova, B; Smetana, K, Jr; Holikova, Z; Broz, L; Gabius, HJ. Mannosides as crutial part of bioactive supports for cultivation of human keratinocytes without feeder cells. *Biomaterials*, 2003 24, 863-872

[136] Pradny, M; Lokaj, J; Novotna, M.; Sevcik, S. Interaction between the amino and carbonyl groups in copoly(2-dimethylaminoethyl methacrylate – N-phenylmaleimide). *Makromol. Chem., Macromol. Chem. Phys.*, 1989 190, 2229–2234.

[137] Pradny, M; Sevcik, S. Precursors of hydrophilic polymers. 3. Behavior of isotactic and atactic poly(2-dimethylaminoethyl methacrylate) in water ethanol solutions. *Makromol. Chem., Macromol. Chem. Phys.*, 1985 186, 111–121.

[138] Xi, FN; Liu, LJ;Wu, Q; Lin, XF. One step construction of biosensor based on chitosan-ionic liquid-horseradish peroxidase biocomposite formed by electrodeposition. *Biosens. Bioelectron.*, 2008 24, 29-34.

[139] Shang, J; Shao, ZZ; Chen, X. Chitosan-based electroactive hydrogel. *Polymer*, 2008 49, 5520-5525.

[140] Zhao, J; Sclabassi, RJ; Sun, MG. Biopotential electrodes based on hydrogel. Conference Information: IEEE 31st Annual Northeast Bioengineering Conference, APR 02-03, 2005 Stevens Inst Technol, Hoboken, NJ. Source: 2005 IEEE 31st Annual Northeast Bioengineering Conference Pages: 69-70.

[141] Moussy, F; Harrison, D; O'Brien, D; Rajotte, R. Performance of subcutaneously implanted needle-type glucose sensors employing a novel trilayer coating. *Anal. Chem.*, 1993 65, 2072–2077.

[142] Moser, I; Jobst, G; Urban, G. Biosensor arrays for simultaneous measurement of glucose, lactate, glutamate and glutamine. *Biosens. Bioelectron.*, 2002 17, 297–302.

[143] Ward, W; Jouse, J; Birck, J; Anderson, E; Jansen, L. A wire-based dual-analyte sensor for glucose and lactate: in vivo and in vitro evaluation. *Diabetes Technol. Ther.*, 2004 6, 389-401.

[144] Kameswaran, N; Cullen, DK; Pfister, BJ; Ranalli, NJ; Huang, JH; Zager, EL; Smith DH. A novel neuroprosthetic interface with the peripheral nervous system using artificially engineered axonal tracts. *Neurol. Res.*, 2008 30, 1063-1067.

[145] Asberg, DP; Inganas, O. PEDOT/PSS hydrogel networks as 3-D enzyme electrodes. International Conference on Science and Technology of Synthetic Metals (ICSM 2002), JUN 29-JUL 05, 2002 Shanghai, Peoples R China, Source: *Synthetic met.*, 2003 137, 1403-1404.

[146] Alexander, M. Basics of electrosurgery and argon plasma coagulation: Use and safety. *Gastroenterol. Nurs.*, 2007 30, 140-140.
[147] Huschak, G; Steen, M; Kaisers, UX. Principles and risks of Electrosurgery. *Anasthesiologie, Intensivmedizin, Notfallmedizin, Schmerztherapie AINS*, 2009 44, 10-13.
[148] Springer, S; Bunk, W. Basics of biosignal analysis of ECG and EEG with chaos theoretical methods. *Neuroendocrinol. Lett.*, 2003 24, 232-235.
[149] Hallstrom, A; Rea, T; Mosesso Jr., V; Cobb, L; Anton, A; Van Ottingham, L; Sayre, M; Christenson, J. The relationship between shocks and survival in out-of-hospital cardiac arrest patients initially found in PEA or asystole. *Resuscitation*, 2007 74, 418-426.
[150] Leonhard Lang, GmbH Archenweg, 56 A-6020 Innsbruck, Austria, www.leonhardlang.com/?id=112, Austria, medical@leonhardlang.at
[151] Hejcl, A; Pradny, M; Michalek, J; Jendelova, P; Sykova, E. Biocompatible hydrogels in spinal cord injury repair. *Physiol. Res.*, 2008 57(Suppl 3), S121–132.
[152] Schmidt, CE; Leach, JB. Neural tissue engineering: strategies for repair and regeneration. *Annu. Rev. Biomed. Eng.*, 2003 5, 293–347.
[153] Pradny, M; Lesny, P; Fiala, J; Vacik, J; Slouf, M; Michalek, J; Sykova, E. Macro porous hydrogels based on 2-hydroxyethyl methacrylate. Part 1. Copolymers of 2-hydroxyethyl methacrylate with methacrylic acid. *Collect. Czech. Chem. Commun.*, 2003 68, 812-822.
[154] Pradny, M; Lesny, P; Smetana Jr, K; Vacik, J; Slouf M; Michalek, J; Sykova, E. Macroporous hydrogels based on 2-hydroxyethyl methacrylate. Part 2. Copolymers with positive and negative charges, polyelectrolyte complexes. *J. Mater. Sci.: Mater. Med.*, 2005 16, 767-773.
[155] Pradny, M; Michalek, J; Lesny, P; Hejcl, A; Vacik, J; Slouf, M; Sykova, E. Macroporous hydrogels based on 2-hydroxyethyl methacrylate. Part 5. Hydrolitically degradable materials. *J. Mater. Sci.: Mater. Med.*, 2006 17, 1357-1366.
[156] Woerly, S; Petrov, P; Sykova, E; Roitbak, T; Simonova, Z; Harvey, AR. Neural tissue formativ within porous hydrogels implanted in brain and spinal cord lesions: ultrastructural, immunohistochemical, and diffusion studies. *Tissue Eng.*, 1999 5, 467–88.
[157] Lesny, P; Pradny, M; Jendelova, P; Michalek, J; Vacik, J; Sykova, E. Macroporous hydrogels based on 2-hydroxyethyl methacrylate. Part 4: Growth of rat bone marrow stromal cells in three-dimensional hydrogels

with positive and negative surface charges and in polyelectrolyte complexes. *J. Mater. Sci.: Mater. Med.*, 2006 17, 829–833.

[158] Michalek, J; Pradny, M; Artyukhov, A; Slouf, M; Vacik, J; Smetana Jr, K. Macroporous hydrogels based on 2-hydroxyethyl methacrylate. Part 3. Hydrogels as carriers for immobilization of proteins. *J. Mater. Sci.: Mater. Med.*, 2005 16, 783-786.

[159] Han, DW; Lee, MS; Park, BJ; Kim, JK; Park, JC. Enhanced neurite outgrowth of rat neural cortical cells on surface-modified film of poly(lactic-co-glycolic acid). *Biotechnol. Lett.*, 2005 27, 53-58.

[160] Lakard, S; Herlem, G; Propper, A; Kastjer, A; Michel, G; Valles-Villarreal N. Adhesion and proliferation of cells on new polymers modified biomaterials. *Bioelectrochemistry*, 2004 62, 19–27.

[161] Hejcl, A; Lesny, P; Pradny, M; Sedy, J; Zamecnik, J; Jendelova, P; Michalek, J; Sykova, E. Macroporous hydrogels based on 2-hydroxyethyl methacrylate. Part 6: 3D hydrogels with positive and negative surface charges and polyelectrolyte complexes in spinal cord injury repair. *J. Mater. Sci.: Mater. Med.*, 2009 20, 1571-1577.

Index

A

acid, 44, 56, 73, 83
acrylate, 9, 49
acrylic acid, 49
acrylonitrile, 54, 78
additives, 10, 23
adhesion, 31, 42, 44, 56, 59, 61, 65, 68, 69, 75
adhesion properties, 69
age, 34, 54
alcohol, 10, 44
alcohols, 9
alkyl methacrylates, 44
amines, 58
amino acids, 5, 72
ammonium, 63
argon, 82
arteriovenous malformation, 79
artery, 78, 79
astigmatism, 43
astrocytes, 69
atmospheric pressure, 30
attachment, 44, 60
Austria, 82
axons, 67, 69

B

background, 37, 38
benzene, 28
benzoyl peroxide, 56
biocompatibility, 49, 50, 76, 77
biodegradable materials, vii
biologically active compounds, 58
biomaterials, 67, 69, 76, 78, 83
biomedical applications, vii, viii
biopolymer, 5, 73
biosensors, 63
blood, 55, 56, 67, 79
blood vessels, 55, 67
Boltzmann constant, 16
bonding, 15, 18
bonds, 1, 4, 5, 7, 8, 18
bone, 82
bone marrow, 82
brain, 38, 69, 82
brain tumor, 38
branching, 9, 33
Brownian motion, 37

C

calcification, 51, 77
calcium, 56
capsule, 49, 50, 76
carbonyl groups, 81
cardiac arrest, 82
carrier, viii, 80
cataract, 49, 51, 76
cataract extraction, 49

catheter, 78
cation, 24
cell, vii, viii, 59, 60, 61, 67, 68, 69, 75
cell culture, vii
central nervous system, 67
chains conformation, 22
chemical properties, vii, 68
chemical reactions, 1
chemical structures, 44, 45
China, 81
clinical trials, 59
CNS, 67, 68, 69
coagulation, 56, 79, 82
coatings, 74
collagen, 53
compatibility, vii
compliance, 35
complications, 50
components, 9, 17, 27, 65
composition, 4, 18, 27, 28, 72, 75
compounds, 42
compressibility, 72
compression, 34, 39
concentration, 3, 4, 5, 6, 7, 12, 13, 14, 17, 18, 19, 20, 21, 22, 23, 24, 25, 27, 31, 32, 65
condensation, 24, 73
connective tissue, 69
connectivity, 4, 12, 16
construction, 50, 81
contrast sensitivity, 51
control, 27, 34, 35, 68, 76
control group, 76
conversion, 3, 4, 5, 6, 7, 8, 12, 29, 30
copolymers, 5, 44, 46, 50, 54
cornea, 41, 42
correlation, 21, 71, 77
covalent bond, 1, 46
critical value, 12
crystalline, 18, 72
crystallites, 1
cultivation, vii, 58, 59, 60, 61, 80, 81
culture, viii, 61
cycles, 7, 33

D

defects, vii, 57, 61
defibrillation, 63, 64, 65
definition, 1, 17
deformation, 16, 30, 31, 32, 33, 34, 35, 50, 73
degradation, 67
delivery, 27, 76
density, 9, 10, 12, 20, 25, 26, 32, 68
developed countries, 43
differentiation, 19, 61, 68
diffusion, 23, 43, 46, 82
diffusivities, 73
diluent, 9, 10, 12, 17, 27, 29, 32, 39, 73
dimethacrylate, 8, 9, 13, 14, 20, 21, 50, 56
displacement, 4, 37
dissociation, 2, 5
distribution, 5, 7, 8, 31
double bonds, 7, 8, 13
dressing material, 57, 58
dressings, 57, 58, 79, 80
drug delivery, 27, 50, 67, 79
drugs, vii, viii
drying, 31, 46, 50, 65

E

economic problem, 57
EEG, 63, 65, 82
elasticity, 16, 31, 33, 37, 68, 72, 74
elastomers, 42, 44, 46, 74
electrodeposition, 81
electrodes, 63, 65, 81
electrolyte, 63
elongation, 31
emboli, 55, 56, 78, 79
embolization, 55, 56, 78, 79
embolus, 56
encapsulation, 54
energy, 30, 35
entanglements, 22, 46
environment, 23, 27, 38, 57, 58, 67
enzymes, 63
epidermis, 80

epithelia, 80
equality, 27
equilibrium, vii, 10, 15, 17, 25, 26, 27, 29, 31, 32, 33, 34, 35, 36, 39, 43, 44, 63, 65, 73
ethanol, 81
ethylene, 10, 13, 14, 20, 21, 50, 78
ethylene glycol, 10, 78
evaporation, 34
evolution, 7, 57

F

fibroblasts, 59
fibrosis, 54
fibrous tissue, 54
France, 41
free radicals, 58
free volume, 24, 31

G

gases, 42, 43
gel, vii, 1, 3, 4, 5, 6, 7, 8, 10, 11, 12, 13, 15, 16, 17, 20, 22, 23, 24, 25, 27, 28, 29, 30, 31, 37, 38, 43, 44, 58, 68, 71, 73, 74, 79
gel formation, 8, 10
gelation, 7, 38, 71, 72
generation, 41, 46
Georgia, 50
Germany, 41
Gibbs energy, 19, 30
glass transition, 24, 31, 49
glass transition temperature, 24, 49
glucose, 81
glutamate, 81
glycerol, 9, 44, 65
glycol, 8, 28, 56, 73
groups, 3, 4, 5, 6, 7, 8, 9, 10, 15, 24, 25, 27, 46, 49, 51, 61, 69, 77
growth, 7, 8, 9, 10, 38, 39, 43, 59, 60, 61, 68, 69, 74

H

haemoptysis, 55, 79
healing, 57, 58, 60, 79, 80

histochemistry, 61
homopolymers, 63
hydrogels, vii, viii, 1, 2, 3, 5, 7, 8, 9, 15, 21, 23, 27, 31, 32, 37, 38, 43, 44, 45, 46, 47, 53, 54, 57, 58, 61, 63, 65, 67, 68, 69, 70, 72, 73, 75, 76, 82, 83
hydrogen, 1, 15, 18
hydrogen bonds, 1
hydrophilicity, vii, 9, 46
hydrophobicity, 20, 42
hydroxyl, 9
hyperbranched polymers, 8
hysteresis, 31

I

ideal, 8, 33, 58
image, 38, 75
image analysis, 38
immobilization, 56, 83
implementation, 51
in transition, 31
in vitro, 61, 81
in vivo, 38, 81
independence, 3, 19
industrialized countries, 50
industry, 42
infection, 58
infinite, 3, 7, 17, 18, 20, 30, 33
inflammation, 80
interaction, viii, 11, 12, 15, 16, 17, 18, 19, 20, 21, 23, 24, 25, 27, 28
interactions, 1, 15, 17, 18, 19, 22, 25, 28, 46, 72
intercellular contacts, 61
interdependence, 17
interface, 81
interrelations, 19, 30
intraocular, 76, 77
iodine, 56
ion exchangers, 24
ionizable groups, 24, 25
ionization, 25, 26, 27
ions, 24, 25, 27, 43
iris, 49
irradiation, 27, 59

K

keratin, 61
keratinocyte, 58, 59, 60, 61
keratinocytes, 59, 60, 61, 80, 81
kinetics, 73

L

lactose, 61
language, 32
lens, vii, 41, 42, 43, 44, 46, 47, 49, 50, 51, 73, 75, 76, 77
lesions, 82
light scattering, 3, 71
linear dependence, 5, 20
liquids, 28, 50
localization, 79

M

macromolecular chains, 32
macromolecules, 72
macrophages, 55
magnetic field, 27, 38
management, 53, 57, 80
manufacturing, 49
market, 42, 43, 45, 46, 57, 58
material surface, 44, 51
materials science, 38
matrix, 10
measurement, 6, 24, 32, 33, 34, 35, 37, 38, 81
mechanical properties, 38, 49, 56, 63, 68, 76
mechanical stress, 38, 41, 42
media, 8, 43, 72
memory, 17, 32
mesenchymal stem cells, 69
metabolism, 41
metabolites, 43, 58
methacrylates, 44
methacrylic acid, 9, 44, 50, 56, 58, 82
methyl methacrylate, 8, 56
mice, 56, 59, 60
microclimate, 59
micrometer, 34

microscope, 38
microspheres, 55, 56, 78, 79
microstructure, 37
migration, 54, 67
mixing, 15, 16, 24
model, 25, 26, 33, 37, 61, 73, 79
modulus, 31, 32, 33, 34, 35, 36, 39, 46
molar volume, 18, 21, 22
mold, 42, 76
molecular biology, 69
molecular structure, 46
molecular weight, 3, 4, 5, 7, 8, 17, 18, 20, 25, 33, 43, 46
molecular weight distribution, 7
molecules, 4, 15, 44, 46
monolayer, 59, 61
monomers, 8, 9, 12, 44, 45, 56, 63, 65, 68
morphology, vii, 10, 67, 68
myopia, 42

N

network, 1, 4, 5, 7, 9, 10, 11, 12, 16, 17, 21, 22, 24, 25, 28, 29, 31, 32, 33, 34, 50, 55, 59, 60, 68, 74
neural network, 69
neural networks, 69
neurons, 39, 67, 69
neuroscience, 67
nonirritant, 58

O

obstruction, 53, 54, 56
oligodendrocytes, 67
oligomers, 10, 46
one dimension, 30
opacification, 51, 76, 77
optical properties, 41
order, 39, 49, 50, 55, 67
osmotic pressure, 16
oxygen, 23, 41, 42, 43, 44, 46, 67

P

parameter, 11, 12, 16, 17, 20, 21, 23, 24, 27, 28, 31, 33

parameters, vii, 17, 18, 27, 28, 34, 42, 49, 68
particles, 38, 53, 54, 56, 63, 71
patents, 42, 43, 46
pathophysiology, 69
peripheral nervous system, 81
permeability, 42, 43, 44, 46, 67
pH, 1, 27, 68
pharmaceuticals, 27
phase diagram, 72
physical interaction, 2, 46
PMMA, 42, 43, 49, 51
poly(2-hydroxyethyl methacrylate), 39, 43, 59, 72, 73
poly(methyl methacrylate), 73
polyacrylamide, 78
polyamides, 9
polybutadiene, 37
polycondensation, 73
polydispersity, 3
polymer, vii, 1, 3, 7, 10, 12, 15, 16, 17, 18, 20, 24, 25, 26, 27, 28, 29, 31, 33, 34, 37, 43, 46, 55, 58, 59, 60, 61, 63, 68, 71, 72, 73, 74, 77, 78, 80
polymer blends, 72
polymer chains, 7
polymer matrix, 31
polymer networks, 33, 37, 72, 73
polymer solutions, 1, 18, 73
polymerization, 7, 8, 10, 12, 13, 37, 51, 56, 65
polymers, vii, 1, 8, 10, 17, 18, 36, 37, 43, 44, 49, 55, 71, 72, 73, 74, 75, 81, 83
polyurethanes, 9
polyvinyl alcohol, 56
population, 50
power, 19, 41, 42, 51, 65
prediction, 24
pressure, 10, 12, 15, 17, 23, 24, 30, 31, 53, 54, 57, 73, 74, 80
prevention, 79, 80
probability, 51
production, 41, 42, 43, 50, 61, 80
proliferation, vii, 38, 59, 60, 61, 68, 83
propagation, 7

prostatectomy, 77
proteins, 59, 83
PVA, 10

Q

quartz, 24

R

radical polymerization, 9, 27
radiopaque, 78
range, vii, 1, 20, 21, 27, 31, 33, 44, 51, 57
reaction mechanism, 5
reaction rate, 13
reaction time, 3
reactive groups, 4
reason, 12, 24, 29
recognition, 5
recovery, 31, 69
refractive index, vii, 49, 50
regenerate, 39
regeneration, 67, 69, 82
region, 4, 7, 12, 24, 28, 29, 33, 55
relationship, 32, 82
relaxation, 1, 35, 39
René Descartes, 41
repair, 58, 82, 83
resistance, 31, 53, 54
retina, 50
rheology, 37
rheometry, 36, 37
roentgen, 55, 56
rubber, 16, 24, 31, 33
rubbery state, 24, 32

S

salts, 44, 63
saturation, 24
scavengers, 58, 80
separation, 8, 10, 11, 12, 13, 14, 15, 22, 29, 30, 31, 72, 74
serum, 59, 60
shape, 1, 24, 41, 42, 49, 50, 55, 56, 68
shear, 32, 35, 36, 38, 39
silicones, 46, 49, 50

skin, 57, 58, 59, 61, 65, 79, 80
social problems, 53
sodium, 50, 54, 65, 69
soft matter, 34
software, 35
solvent molecules, 15, 16, 25
solvents, 18, 23, 65
sorption, 24, 31
space, 30, 41
species, 42, 43, 46
spectroscopy, 36, 37, 78
sphincter, 53, 77
spinal cord, 67, 69, 82, 83
spinal cord injury, 67, 82, 83
stability, 10, 29
stoichiometry, 9
storage, 36, 65
strain, 31, 32, 33, 34, 35, 36, 37
stress, 31, 32, 33, 34, 35, 37, 42, 53, 74, 77, 78
stroma, 49
stromal cells, 82
structural changes, 3
structure formation, 71
surface modification, 46
surgical intervention, 51
survival, 82
suture, viii
swelling, vii, 2, 10, 12, 13, 14, 15, 16, 17, 18, 19, 20, 21, 22, 23, 24, 25, 26, 27, 28, 29, 30, 31, 32, 43, 46, 50, 53, 54, 63, 73, 74
Switzerland, 41, 74
synthesis, 7, 71
synthetic polymers, 42

T

temperature, 1, 7, 10, 11, 15, 16, 17, 18, 19, 20, 27, 32, 34, 43, 50, 72, 73
temperature dependence, 17, 19
therapeutic process, 57
therapy, vii, 61, 67
thermodynamics, 15, 73
thrombin, 56, 79

tissue, vii, viii, 39, 41, 43, 50, 57, 58, 59, 63, 67, 68, 69, 75, 80, 82
torsion, 34, 35
tradition, 59
transformation, 4
transition, 2, 20, 22, 23, 24, 25, 27, 71, 73
transition temperature, 73
transitions, 27, 71, 73
transplantation, vii, 59
transport, vii, 43, 72
trauma, 58
tumor, 38, 74
tumors, 55

U

UK, 50, 57
uniform, 46, 56
universal gas constant, 32
urethra, 53
urinary bladder, 54
urine, 53, 54
uterine fibroids, 55, 78
UV, 50, 65
UV light, 50
UV radiation, 65
uveitis, 76

V

vapor, 23, 29, 31
vesicle, 53
vessels, 55
vinyl monomers, 9
viscosity, 38
vision, 47, 49, 51
vitamins, 58

W

water vapor, 73
weak interaction, 15
wettability, 43, 46
women, 54, 78
World War I, 42
wound healing, 57, 58, 59